U.S. SENATE

COMMITTEE ON COMMERCE, SCIENCE, AND TRANSPORTATION

HON. JOHN MCCAIN, Arizona, *Chairperson*

TED STEVENS, Alaska
CONRAD BURNS, Montana
TRENT LOTT, Mississippi
KAY BAILEY HUTCHISON, Texas
OLYMPIA J. SNOWE, Maine
SAM BROWNBACK, Kansas
GORDON SMITH, Oregon
PETER G. FITZGERALD, Illinois
JOHN ENSIGN, Nevada
GEORGE ALLEN, Virginia
JOHN O. SUNUNU, New Hampshire

ERNEST F. HOLLINGS, South Carolina
DANIEL K. INOUYE, Hawaii
JOHN D. ROCKEFELLER IV, West Virginia
JOHN F. KERRY, Massachusetts
JOHN B. BREAUX, Louisiana
BYRON L. DORGAN, North Dakota
RON WYDEN, Oregon
BARBARA BOXER, California
BILL NELSON, Florida
MARIA CANTWELL, Washington
FRANK LAUTENBERG, New Jersey

FLOYD DESCHAMPS, *Republican Senior Professional Staff*
KEN LA SALA, *Republican Professional Staff*
JEAN TOAL EISEN, *Democratic Senior Professional Staff*

CONTENTS

February 12, 2003

Page

Opening Statements

Witness

SPACE SHUTTLE COLUMBIA

JOINT HEARING

BEFORE THE

SUBCOMMITTEE ON SPACE AND AERONAUTICS

COMMITTEE ON SCIENCE

HOUSE OF REPRESENTATIVES

AND THE

COMMITTEE ON COMMERCE, SCIENCE, AND TRANSPORTATION

U.S. SENATE

ONE HUNDRED EIGHTH CONGRESS

FIRST SESSION

FEBRUARY 12, 2003

Serial No. 108–2

Printed for the use of the House Committee on Science and Senate Committee on Commerce, Science, and Transportation

Available via the World Wide Web: http://www.house.gov/science

U.S. GOVERNMENT PRINTING OFFICE

85–090PS WASHINGTON : 2003

For sale by the Superintendent of Documents, U.S. Government Printing Office
Internet: bookstore.gpo.gov Phone: toll free (866) 512–1800; DC area (202) 512–1800
Fax: (202) 512–2250 Mail: Stop SSOP, Washington, DC 20402–0001

COMMITTEE ON SCIENCE

HON. SHERWOOD L. BOEHLERT, New York, *Chairman*

LAMAR S. SMITH, Texas
CURT WELDON, Pennsylvania
DANA ROHRABACHER, California
JOE BARTON, Texas
KEN CALVERT, California
NICK SMITH, Michigan
ROSCOE G. BARTLETT, Maryland
VERNON J. EHLERS, Michigan
GIL GUTKNECHT, Minnesota
GEORGE R. NETHERCUTT, JR., Washington
FRANK D. LUCAS, Oklahoma
JUDY BIGGERT, Illinois
WAYNE T. GILCHREST, Maryland
W. TODD AKIN, Missouri
TIMOTHY V. JOHNSON, Illinois
MELISSA A. HART, Pennsylvania
JOHN SULLIVAN, Oklahoma
J. RANDY FORBES, Virginia
PHIL GINGREY, Georgia
ROB BISHOP, Utah
MICHAEL C. BURGESS, Texas
JO BONNER, Alabama
TOM FEENEY, Florida
VACANCY

RALPH M. HALL, Texas
BART GORDON, Tennessee
JERRY F. COSTELLO, Illinois
EDDIE BERNICE JOHNSON, Texas
LYNN C. WOOLSEY, California
NICK LAMPSON, Texas
JOHN B. LARSON, Connecticut
MARK UDALL, Colorado
DAVID WU, Oregon
MICHAEL M. HONDA, California
CHRIS BELL, Texas
BRAD MILLER, North Carolina
LINCOLN DAVIS, Tennessee
SHEILA JACKSON LEE, Texas
ZOE LOFGREN, California
BRAD SHERMAN, California
BRIAN BAIRD, Washington
DENNIS MOORE, Kansas
ANTHONY D. WEINER, New York
JIM MATHESON, Utah
DENNIS A. CARDOZA, California
VACANCY
VACANCY
VACANCY

SUBCOMMITTEE ON SPACE AND AERONAUTICS

DANA ROHRABACHER, California, *Chairman*

LAMAR S. SMITH, Texas
CURT WELDON, Pennsylvania
JOE BARTON, Texas
KEN CALVERT, California
ROSCOE G. BARTLETT, Maryland
GEORGE R. NETHERCUTT, JR., Washington
FRANK D. LUCAS, Oklahoma
JOHN SULLIVAN, Oklahoma
J. RANDY FORBES, Virginia
ROB BISHOP, Utah
MICHAEL BURGESS, Texas
JO BONNER, Alabama
TOM FEENEY, Florida
SHERWOOD L. BOEHLERT, New York

BART GORDON, Tennessee
JOHN B. LARSON, Connecticut
CHRIS BELL, Texas
NICK LAMPSON, Texas
MARK UDALL, Colorado
DAVID WU, Oregon
EDDIE BERNICE JOHNSON, Texas
SHEILA JACKSON LEE, Texas
BRAD SHERMAN, California
DENNIS MOORE, Kansas
ANTHONY D. WEINER, New York
VACANCY
RALPH M. HALL, Texas

BILL ADKINS *Subcommittee Staff Director*
ED FEDDEMAN *Professional Staff Member*
RUBEN VAN MITCHELL *Professional Staff Member*
KEN MONROE, *Professional Staff Member*
CHRIS SHANK *Professional Staff Member*
RICHARD OBERMANN *Democratic Professional Staff Member*
TOM HAMMOND *Staff Assistant*

Appendix 1: Answers to Post-Hearing Questions

Appendix 2: Additional Material for the Record

SPACE SHUTTLE COLUMBIA

FEBRUARY 12, 2003

SUBCOMMITTEE ON SPACE AND AERONAUTICS,
HOUSE OF REPRESENTATIVES,
COMMITTEE ON SCIENCE,
JOINT WITH U.S. SENATE,
COMMITTEE ON COMMERCE, SCIENCE,
AND TRANSPORTATION,
Washington, DC.

The Committees met, pursuant to notice, at 9:35 a.m. in Room SR–325, Russell Senate Office Building, Hon. John McCain, Chairman of the Senate Committee, presiding.

OPENING STATEMENT OF HON. JOHN McCAIN, U.S. SENATOR FROM ARIZONA

Chairman MCCAIN. Good morning. I welcome my colleagues from the House Science Committee and Administrator O'Keefe.

To keep this hearing to a reasonable length, I appreciate my colleagues' indulgence in limiting opening statements only to those of the chairmen and Ranking Members of the Senate Committee on Commerce, Science, and Transportation and the House Science Committee.

Following Administrator O'Keefe's statement, all Members will be recognized for four minutes to ask questions. We will alternate between Senators and House Members for questions, which is the normal procedure in joint hearings of this nature.

On February 1st, the Nation suffered a devastating loss. As the Space Shuttle *Columbia* descended from orbit, it broke apart. Debris from the accident is still being collected by government agencies and volunteers with the hope that this evidence will help determine the cause of the accident.

The Space Shuttle crew was a remarkable team of professionals. They were and will always be role models for all Americans. Their dedicated service and sacrifice to promote scientific research not only for our country, but for the world, will never be forgotten. They paid the ultimate price in pursuit of not only their dreams, but the dreams of nations. For that, we will be forever grateful.

As we look to the future of the space program, we can pay tribute to our fallen heroes by diligently carrying out our responsibilities as legislators. In today's hearing, we hope to examine what went wrong on February 1st, the status of the investigation, and how we can ensure that an accident like this will never happen again.

This will be the first of a series of hearings on NASA and our space program. While today we're focusing on the *Columbia*, the accident also has focused our attention on the broader policy issues that perhaps we have neglected for too long. In subsequent hearings we will address the role of manned and unmanned space exploration, the costs and benefits of continuing the shuttle program, and our investment in the International Space Station and the effectiveness of NASA management. More fundamentally, we must examine the goals of our space program.

I firmly believe that manned space exploration should continue. Its nature, however, should be and will be examined. We also must examine the extent to which Congress and the Administration may have neglected the Shuttle's safety program. A comprehensive examination necessitates a review of our own actions and those of the Administration to determine if the Shuttle program was underfunded or managed in a manner that compromised safety.

I applaud Administrator Sean O'Keefe and NASA for their openness and availability. This has been an extraordinarily trying time for everyone in the agency. The Administrator and other officials have conducted themselves in a manner worthy of an agency that is not only a national brain trust, but is entrusted with realizing the dreams of all humanity. Many have noted the vast improvement of the release of information, as compared to the *Challenger* tragedy of 1986.

I would like to assure the families of the brave men and women who died aboard the *Columbia* and the dedicated employees of NASA that we will do everything in our power to identify the cause of this tragedy and remedy it.

I thank Administrator O'Keefe and his team for appearing before us today, and I look forward to the testimony.

STATEMENT OF HON. SHERWOOD L. BOEHLERT, U.S. REPRESENTATIVE FROM NEW YORK

Chairman BOEHLERT. We usually open hearings by talking about what a pleasure it is to be here today. But, of course, that is not the case. I'm reminded of what Lyndon Johnson said when he appeared before Congress after the Kennedy assassination. He said, "All I have, I gladly would have given not to be standing before you today." I'm sure that is the way we all feel with the tragic loss of the *Columbia* crew so fresh in our minds and in our hearts.

But we owe it to those astronauts and their families and to the American public to work as hard as humanly possible to determine the cause of the Shuttle's breakup and to rigorously pursue all the policy questions the accident brings to a head.

I view this hearing as a start of a very long conversation we will all be having about the *Columbia* incident and its ramifications. I think that it's very appropriate that we start that conversation on a bicameral basis, and I want to thank Senator McCain for being so willing to make this a Joint Hearing. The House and the Senate and NASA are going to have to cooperate as we each review the accident and the human space flight program, and our joint work today should send a clear signal that we can and will do just that.

We will also be coordinating with the *Columbia* Accident Investigation Board headed by Admiral Gehman. I've spoken to Admiral Gehman, and I am impressed with the Admiral's determination to be independent and deliberate, vowing to be swayed neither by outside pressures or artificial deadlines. And I appreciate the swiftness with which Administrator O'Keefe activated the board.

That said, the more I've read the board's charter, the more I've become convinced that it must be rewritten. The words of the charter simply do not guarantee the independence and latitude that both the Administrator and the Admiral have sincerely promised. The charter's words need to match everyone's intent now to avoid any problems later. I also continue to believe that several more members should be added to the board to ensure that it has the appropriate breadth of experience and expertise.

We will be working closely with the board as the Science Committee proceeds with its own bipartisan investigation, which will focus on the many policy questions raised by the accident. We're going to have to raise some tough and basic questions that have gone unanswered for too long.

What are the true risks of flying the Shuttle, especially if it's going to remain in service for another 10 to 15 years? What are the true costs of continuing the Shuttle program at specific levels of risk? And what are the advantages of investing in the Shuttle, as compared to investing in other NASA programs, other R&D programs, and, indeed, other government programs, in general?

But we can't begin to deal with those overarching issues until we have a better sense of what happened to the *Columbia* and why, and it's obviously too soon to expect to know that.

No one should expect any revelations at today's hearing. We are here today to get a status report. We ought to avoid pronouncements today that we may later come to regret.

I'm reminded of an interview I once read with an executive of the utility that owned Three Mile Island at the time of the accident there. He was asked, "What was the worst thing you did in handling the accident?" He answered immediately. He said, "We just didn't have the presence of mind to say, 'I don't know.'"

I would advise Administrator O'Keefe, who has responded magnificently in this time of crisis, don't hesitate to say, "I don't know." You're still in search of elusive answers.

Despite the best of intentions, NASA has at times already put out misleading information because it didn't check the facts. For example, information indicating that environmental rules could have contributed to the accident have so far turned out to be entirely spurious, but it's taken NASA a long time to clarify its statements.

Today is a chance to put facts into the record, facts that will help chart NASA's future. If we are to find the facts and honor the memory of the *Columbia* crew, we have to approach our task in a true spirit of exploration, with open and probing minds, without preconceived notions or foregone conclusions. That should be our goal today.

Thank you, Mr. Chairman.

[The prepared statement of Chairman Boehlert follows:]

PREPARED STATEMENT OF CHAIRMAN SHERWOOD L. BOEHLERT

We usually open hearings by talking about what a pleasure it is to be here. But of course today that is not the case. I'm reminded of what Lyndon Johnson said when he appeared before Congress after the Kennedy assassination: "All I have, I gladly would have given, not to be standing before you today." I'm sure that is the way we all feel, with the tragic loss of the *Columbia* crew so fresh in our minds and in our hearts.

But we owe it to those astronauts and their families, and to the American public, to work as hard as is humanly possible to determine the cause of the Shuttle's breakup and to rigorously pursue all the policy questions the accident brings to a head.

I view this hearing as the start of a very long conversation we will all be having about the *Columbia* incident and its ramifications. I think that it's very appropriate that we start that conversation on a bicameral basis, and I want to thank Senator McCain for being so open to making this a joint hearing. The House and the Senate and NASA are going to have to cooperate as we each review the accident and the Human Space Flight program, and our joint work today should send a clear signal that we can and will do just that.

We will also all be coordinating with the *Columbia* Accident Investigation Board, headed by Admiral Gehman. I spoke to Admiral Gehman earlier this week, as did our Committee staff on a bipartisan basis. I am impressed with the Admiral's determination to be independent and deliberate, vowing to be swayed neither by outside pressures or artificial deadlines. That's the right attitude, and we will be watching to ensure that it guides the Board's proceedings.

We will be working with Admiral Gehman as the Science Committee proceeds with its own bipartisan investigation, which will focus on the many policy questions raised by the accident. In the end, we must have a full appraisal and open debate about the true risks of flying the Shuttle, the true costs of continuing the Shuttle program at specific levels of risk, and the comparative advantages of investing in the Shuttle as opposed to other NASA programs, or indeed as opposed to other R&D programs or government programs, in general.

But we can't begin to deal with those overarching issues until we have a better sense of what happened to the *Columbia* and why, and it's obviously too soon to expect to know that. No one should expect any revelations at today's hearing. We are here today to get a status report.

We all ought to avoid pronouncements today that we may later come to regret. I'm reminded of an interview I once read with an executive of the utility that owned Three Mile Island at the time of the accident there. He was asked, "What was the worst thing you did in handling the accident?" He answered immediately. He said, "We just didn't have the presence of mind to say, 'I don't know.'" I would advise Administrator O'Keefe, who has responded magnificently in this time of crisis: don't hesitate to say, "I don't know."

Despite the best of intentions, NASA has already sometimes put out misleading information because it didn't check the facts. For example, information indicating that environmental rules could have contributed to the accident has so far turned out to be entirely spurious. But it's taken NASA a long time to clarify its statements.

One reason I believe that today's hearing can be useful is that with so much information is already floating around from so many sources, it's important that Congress and NASA have an opportunity to create a clear record of where things stand at this point.

It's especially important today that we get a clear sense of how NASA will handle the investigation and what contingency plans are in place should the Shuttle be grounded for an extended period of time. I'm sure we will also examine how NASA had been viewing the long-range safety of the Shuttle prior to the accident and how this may already have changed.

All of us are still mourning the loss of the *Columbia* crew, but we must now turn to planning the future. And we can do that only in a true spirit of exploration—with a full and open examination of all the facts, without preconceived notions or foregone conclusions. That process starts today. Thank you.

Chairman MCCAIN. Senator Hollings.

STATEMENT OF HON. ERNEST O. HOLLINGS, U.S. SENATOR FROM SOUTH CAROLINA

Senator HOLLINGS. Thank you very much, Mr. Chairman, for calling this meeting. We welcome our colleagues from the House side and my old friend, Administrator O'Keefe. We're glad to have you with us.

Chairman Boehlert, I got the message, "Let's don't jump to conclusions." I'm reading in the morning paper a similar message—Admiral Gehman said that "the investigation with solid evidence thus far hard to come by." On the contrary, we have a lot of solid evidence that we've come by, and I sort of discern some kind of eery avoidance here of what really happened.

Here is the hard evidence. NASA's had a long history of problems with the Shuttle's heat tiles. We know that the *Columbia*'s VERY first mission, many of the tiles flew off. That's 22 years ago. In 1994, a study entitled the *Risk Management for the Tiles of the Space Shuttle,* by Stanford and Carnegie Mellon University, found that 15 percent of the Shuttle's tiles account for 85 percent of the risk. And that was confirmed by a 1997 study by the National Academy of Sciences.

Then a study by the Johnson Space Center in March of 2000 found that the leading edges of the wing, quote, "pose the highest risk for critical failure," end quote. And then during the launch of the *Columbia* on January the 16th, we have video evidence of debris striking the Shuttle orbiter 81 seconds after launch, potentially causing a gash in the left wing of some 30 inches long, seven inches wide, and over two-and-a-half inches thick. And then 18 minutes from landing, the Shuttle was pitching and yawing due to drag on the left wing.

And, of course, this morning's paper says that as it was coming down, and I'm quoting the Mission Control, "FYI," for your information, "I've just lost four separate temperature transducers on the left side of the vehicle, hydraulic return temperatures," he calmly reported.

Again quoting, eight minutes before all communications was lost, Mr. Kling noticed the loss of data from temperature gauges on the left wing on the spacecraft as he monitored the Shuttle's descent into the atmosphere. A few moments later, Mr. Kling reported drag on the spaceship, but controllers expressed no alarm.

And, finally, the elevons, the picture that showed the elevons, tried to counteract that drag in engine thrusters to help it gain control, because a minute before the explosion, the U.S. Air Force captured that picture of the Shuttle showing a bulge of deformity along the front edge of the left wing. Right on down the list.

Mr. Administrator, I would think, in the testimony, we all agree that we don't want to jump to conclusions. We all agree to be very thorough and leave no stone unturned, but we do have a rebuttable presumption here that the damage to the left wing at the time of liftoff was the cause. And let's rebut it. Let's find something. But don't all of a sudden be discovering debris all around and all of these other things that pant one way and say we have no idea what happened.

I have been in these investigations before, and we knew exactly what happened at the *Challenger*. Allen McDonald said he was in the control room. They had warned not to take it off. It was going to cause a catastrophe. And he said one gentlemen said, "There she goes." Another one said, "Like a piece of cake." Then he said it exploded, and everyone in the room knew why. We spent years investigating to find out the same thing that we knew immediately at the time of explosion.

Thank you, Mr. Chairman.

[The prepared statement of Senator Hollings follows:]

PREPARED STATEMENT OF SENATOR ERNEST O. HOLLINGS

I would like to begin by offering my condolences to the family members of the Space Shuttle *Columbia* crew of mission STS–107. These heroes gave their lives in the advancement of science and all Americans should be overwhelmed by their sacrifice. The *Columbia* crew was on a special mission to conduct scientific research in outer space. As a strong supporter of scientific research, I'm grateful to all the men and women of NASA who undertake such endeavors to advance scientific knowledge.

Welcome Administrator O'Keefe. You are here today to provide my distinguished colleagues and I with answers of how this tragedy was allowed to happen. There is a question as to whether this committee has consistently provided NASA with the funds it has requested for the Space Shuttle program. We want to get to the bottom of this accident so that we can ensure that it does not happen again.

Now I know that the NASA engineers have developed this "fault tree" to identify all the possible causes of this tragic event. Branches are continually added, but nothing is eliminated. Investigators are exploring every lead, but the facts of the matter are:

- We have video evidence of debris striking the Shuttle orbiter 81 seconds after launch. Engineers estimated the damaged tile area in the *left wing* to be 30 inches long by 7 inches wide, yet there was no concern for the tiles failing upon re-entry into the Earth's atmosphere.
- NASA's had a long history of problems with the heat tiles. These problems date back to 1981 when the first *Columbia* launch came back with lost or damaged tiles.
- NASA has recognized the tile problem. Numerous studies have been conducted. In 1990 a study found that *15 percent of the Shuttle's tiles account for 85 percent of its risks* and recommended that improving maintenance procedures could reduce the probability of tile related Shuttle accidents by 70 percent.
- Less than 18 minutes from landing, the Shuttle was pitching and yawing due to drag on the *left wing*. Its elevons tried to counteract the drag and engine thrusters had fired to gain control.

It is clear that we have a rebuttable presumption to go forward with the investigation to focus the examination on how the tiles failed causing the catastrophic failure.

Chairman MCCAIN. Thank you, Senator Hollings.

Congressman Hall hasn't arrived yet, so we will proceed to Mr. O'Keefe, the Administrator of the National Aeronautics and Space Administration. He's accompanied by Mr. Frederick D. Gregory, who is the Deputy Administrator, and Mr. William Reedy, the Associate Administrator for Space Flight. If you'd like to join—or they can remain where they are.

And, again, I want to thank you for the extreme willingness on your part to share all information that you have with not only Members of Congress, but with the American people.

Please proceed, Mr. O'Keefe, and I hope you understand that we're interested in as thorough a briefing as possible, as are Americans who are viewing this hearing today.

Thank you.

[The prepared statement of Senator Lautenberg follows:]

PREPARED STATEMENT OF SENATOR FRANK R. LAUTENBERG

Mr. Chairman,

Today's hearing on the Space Shuttle *Columbia* disaster and the hearings likely to follow in the weeks and months ahead will bring additional pain to that which we already feel while in a period of mourning for seven brave, exceptional human beings in the prime of life. The hearings will also bring pain because, frankly, indications are that some earlier warnings might have raised questions about whether or not presumption of risk was insufficiently reviewed.

The Space Shuttle *Columbia* disaster forces us to ask difficult questions. The Federal Government has spent more than $60 billion on the Space Shuttle program, the International Space Station, and the X–33/VentureStar Space Plane (which advocates believed would replace the Shuttle). Our fleet of Shuttles is grounded at least until we determine what caused the *Columbia* accident and fix it; the three-person crew of the Space Station spends 80 percent of their time on maintenance; and the Bush Administration has canceled the Space Plane project. As a result of that cancellation, we now intend to continue using Shuttles at least until 2012, and possibly beyond 2020. Some of the technology on the Shuttles is 30 years old. We never intended to use them this long.

I want to make it clear that I feel that the Shuttle astronauts made a major contribution to our effort to assess the value to humankind of research in space, and I grieve over their deaths. The desire to reach for the stars is as old as human history and the ambitions embodied in our manned space program are noble ones. But we have had two fatal accidents in 113 Shuttle missions. Many people have become inured to the dangers inherent in sending people into space and bringing them back safely. But the fact is, it's a high-risk venture. Some risk is unavoidable—that's what makes our astronauts such brave individuals. But are we willing to divert precious resources available for other essential research and experimentation planned or in place to reduce the risks of manned space exploration to the point where they become acceptable?

Because of the downturn in the economy that started in March 2001, the September 11th terrorist attacks, and the tax cuts enacted that year, we are facing federal budget deficits "as far as the eye can see." And now the Administration proposes to reduce federal revenues even more. How can we guarantee that we can spend what it takes to make the space effort safer and successful? If we make the investment necessary, what benefits will we reap from continued Shuttle operations? What are the "opportunity costs" of such an investment? In other words, what other national priorities will suffer in the battle for scarce funds? Our manned space exploration program has been long on ambition but increasingly short on the hard-headed assessments needed to answer these fundamental questions.

Manned space exploration isn't cheap. If we try to do it on the cheap, we put safety—and people's lives—at risk. I'm sure we will hear in testimony today and in the future that safety has never been compromised. But NASA has always had problems overseeing its contractors. And the National Research Council has concluded that the contract to manage the Shuttle program awarded to United Space Alliance in 1996 contained financial incentives for investments in efficiency, but not for investments in modernization and safety improvements.

Much of today's hearing and the hearings to come will focus on technical matters—possible causes of the *Columbia* accident, possible safety improvements. I am interested to know, for instance, what steps—if any—NASA took to ensure *Columbia*'s safe re-entry after determining that debris—presumably foam insulation from the fuel tank—hit and may have damaged the left win during lift-off. I am also interested in learning from NASA Administrator Sean O'Keefe what additional safety precautions might have been assured with greater funding. And I want to know what safety upgrades, if any, were made after the *Columbia* space flights scheduled for August 2000 and March 2002 were postponed.

In the course of today's hearing and future hearings, we will also scrutinize NASA's relationship with its contractors. We will also review Congress's relationship with NASA. We will analyze Administration budget requests for NASA past and present.

I hope our investigation will be more about fixing problems than fixing blame—although determining accountability obviously is important. But beyond such immediate concerns, I hope we will address the harder question about whether the benefits outweigh the risks when we send people into space at this time and in the current fashion when unmanned missions can almost entirely match the quality of human participation.

[The prepared statement of Ms. Jackson Lee follows:]

PREPARED STATEMENT OF REPRESENTATIVE SHEILA JACKSON LEE

Mr. Chairman,

Thank you for calling this hearing and bringing us all together to speak and learn about the *Columbia* tragedy. This is a tough time for all of us from the Houston community, but especially for the team at Johnson Space Center. To the world those astronauts were valiant heroes; to us they were also friends, neighbors, and family—or as the Houston Chronicle proclaimed them, "The Heroes Next Door." I am impressed by the diligence, progress, and openness of the NASA investigators that we have all been getting to know through the press.

Those investigators have a difficult job ahead, and it is essential that that job be done well. We must find all the available facts, and we must not jump to hasty conclusions. It seems that the data is pouring in, in the form of video, computer analysis and collection of debris. I am concerned by reports of loose foam or ice that may have damaged the left wing during liftoff, especially since this may have been a problem in a past mission. I want to know what was done to keep such chunks from detaching and striking our multi-billion dollar Shuttle, entrusted with the lives of 7 Americans.

However, we cannot be myopic and disregard or short-change other evidence and explanations. The inquiries must be methodical and objective. The team must leave no room for suspicion of cover-up or sloppiness. The families of the seven valiant crew members that lost loved ones deserve to know why this tragedy happened, as do the American people. Most importantly, we owe it to our brave future astronauts to show them our commitment to their safety.

I am pleased that after we Democrats in the Science Committee sent a letter to the President expressing our concerns about the independence of the investigatory board, that the hearing and make-up of the board were changed. However, I feel there is still room for improvement. I recommend the inclusion of Nobel Laureates, academicians, and depending on their interests—perhaps family members of lost crew. It is important that the team is weighted toward bright people, who are not employees of NASA, and who do not have close personal ties to NASA or the Administrator.

The conclusions we all reach must not only be in the form of, "Part A broke, and part B got too hot." We must discern what were the factors that led to those parts being included in a vehicle entrusted with seven lives and such an important mission. What were the quality assurance protocols? Were corners cut?

Furthermore, this investigation needs to be expeditious. We have three Space Shuttles with critical missions already planned. We also have the International Space Station, with three astronauts high up above us waiting to hear their own futures. Thankfully, we have partnered with our Russian allies and others and ensured that we have the means to get those astronauts home, even though we may need to ground our own fleet for some time. However, we cannot continually place American lives in the hands of another nation for long. Nor can we risk losing the use of the International Space Station that we have been working so hard, and investing so much, to achieve.

[The prepared statement of Ms. Lofgren follows:]

PREPARED STATEMENT OF REPRESENTATIVE ZOE LOFGREN

I'd like to thank Congressman Boehlert and Senator McCain for convening this hearing. Over the next few months, we will be asking some tough questions related to the breakup and loss of the *Columbia,* and the future of the United States space program. But first, our country has paused to reflect on the heroism of the seven astronauts who gave their lives so that the dreams of humans reaching for the stars can live forever. My thoughts go out to the families of our fallen, and to the extended NASA family.

I am pleased the NASA Administrator Sean O'Keefe has joined us here today. I look forward to hearing from and working with you and the dedicated and hard working members of the NASA employee family, as we seek answers to our concerns about the future of the United States space program. I trust that you will ask us for help, keep us informed and be prepared to make your recommendations to this committee that will help us be able to move our space program forward. I firmly believe this committee must focus on asking the difficult questions that relate to how we are best able to resume our quest to explore space.

This committee must work in a nonpartisan manner and should not waste any time in trying to assess blame or create excuses for things that should have been done to help prevent this immense tragedy and loss. To do so would be a waste of time and money and, more importantly, would dishonor the sacrifices made of the brave *Columbia* crew and devalue the efforts being made by all who seek to ensure that this never happens again.

I believe that our pursuit of answers to this tragedy would best be served by the appointment of a truly independent board of inquiry, much like President Reagan appointed after the *Challenger* disaster. Until that happens, Mr. O'Keefe, I am pleased that you accepted some of the recommendations contained in a letter sent to the President last week by 16 Democratic members of the House Science Committee. I am sure many of our Republican colleagues would have joined us in expressing our concerns about the composition of the review board, and I am confident they would have echoed our concerns. Without these changes, I believe the results of this work would have been viewed with great skepticism and certainly would have suffered without the added, independent expertise of the new members of the board. Just as *Columbia*'s crew went into space seeking to expand our knowledge of space, we must do all in our power to ensure that our investigations will answer more questions than they create.

Mr. Chairman, I am committed to sending humans into space. We are explorers by nature, and I believe we must explore our own planet and those beyond. I believe these hearings need to focus not only on investigating the policy concerns that led to the Shuttle tragedy, but where we go from here in the exploration of space.

- Has NASA shifted monies to the ISS and away from the Shuttle program?
- Are we going to develop the next generation of space vehicle, and should we pursue a single-stage-to-orbit program?
- Should we also develop the use of expendable rockets to ferry equipment and personnel to the International Space Station?
- Are we prepared to fund this program—as I think we should—in the current budget climate?

With this in mind, I believe this committee can best honor the memory of *Columbia*'s crew by conducting an honest examination of the role, if any, of recent budget cuts played in this disaster. Should we take this opportunity to acknowledge that the Space Shuttle has never lived up to its dreams of being a cost effective way of traveling to space? Or are we better served by pursuing a new generation of space vehicles, one that can take advantage of the tremendous advances in our knowledge and our technologies than those present in the remaining Shuttle fleet?

STATEMENT OF SEAN O'KEEFE, ADMINISTRATOR, NATIONAL AERONAUTICS AND SPACE ADMINISTRATION; ACCOMPANIED BY FREDERICK D. GREGORY, DEPUTY ADMINISTRATOR, AND WILLIAM O. READDY, ASSOCIATE ADMINISTRATOR FOR SPACE FLIGHT

Mr. O'KEEFE. Good morning. Thank you, Mr. Chairman, Chairman Boehlert.

I appreciate the opportunity to appear before this hearing of the Senate Commerce, Science, and Transportation Committee and the House Science Subcommittee on Space and Aeronautics to discuss the tragic loss of the courageous crew of the Space Shuttle *Columbia*——

Chairman MCCAIN. Could you pull the microphone a little closer?

Mr. O'KEEFE. —the ongoing investigation into this tragedy and the implications of the loss of *Columbia* to the Nation's space exploration efforts.

This morning, 11 days after the accident, our work continues to honor the solemn pledge we made to the astronauts' families and to the American people, that we'll find out what caused the loss of the *Columbia* and its crew, correct what problems we find, and do our utmost to make sure this never happens again.

We welcome the Joint Committee's interest in working with NASA to determine how we can learn from this tragic accident so that we continue advancing the Nation's research and exploration objectives in space while at the same time striving to ensure that we make human space flight as safe as possible.

Throughout NASA's 45 years of serving the public interest, Congress has been our partner helping us achieve the goals outlined in NASA's congressionally authorized charter. This charter compels NASA to explore, use, and enable the development of space for human enterprise; advance scientific knowledge and understanding of the Earth, the solar system, and the universe; and use the environment of space for research; research, develop, verify, and transfer advanced aeronautics, space, and related technologies.

With the support of Congress, NASA has amassed a record of significant achievements that have tangibly improved the lives of all Americans. And when we have erred, you have helped us right our course.

This morning, you'll be asking tough questions, and that's as it should be. Believe me, none of the questions that you will ask can be any tougher than those we're asking of ourselves. I can assure you, however, that whatever determinations are reached regarding the cause of the accident, you'll find that complacency is not one of them.

An ethos of safety is evident throughout the agency. For example, last year we temporarily halted Shuttle flight operations when tiny cracks of less than two inches were discovered in metal liners used to direct the fuel flow inside the propellent lines on two separate orbiters. We did not fly again until that problem was corrected. In a signal of our continuing commitment to rewarding such diligence, we also made it a point to praise a very young examiner, a fellow named David Strait, the young contract employee who had actually discovered the cracks.

Other flight decisions made throughout the year were made with the goal of operational safety being paramount. And from working with the dedicated employees who keep the Shuttle flying safely, I know they have the utmost regard for the enormity of that duty.

This week, at NASA centers throughout the country and in the field, with the support of more than 2,000 people from more than 20 federal agencies, state and local organizations, the important work of data analysis and recovery operations is continuing. We should all be extremely proud of the work that's being conducted by these dedicated public servants.

President Bush observed last week, "The people of NASA are being tested once again. In your grief, you are responding as your friends would have wished, with focus, professionalism, and unbroken faith in the mission of this agency. Captain Brown was correct, America's space program will go on." We intend to maintain that professionalism he referred to until we reach conclusion and beyond.

This morning, to help frame our discussion, I'd like to review for you the significant actions NASA has taken since the morning of the accident in accord with our contingency plan. In addition to articulating notification of first-response procedures defining the roles and responsibility of mishap response and Mishap Investigation Teams, the plan specifies selections of persons outside of NASA to head an independent, seasoned, accident investigation team. Now, while we did not foresee this tragedy, our response has unfolded as we had planned and prepared for in that contingency plan that we had hoped to never have to activate.

This plan was one of many positive outcomes from the terrible loss of the Space Shuttle *Challenger* 17 years ago. So we developed the plan shortly after that and have updated it before every flight.

And a contingency was simulated for this very event just three months ago.

When we first became aware of the problems with STS–107, I was waiting at the Space Shuttle landing strip at the Kennedy Space Center, Cape Canaveral, on Saturday morning, February the 1st. At 8:59 a.m. eastern time, we lost communication with the *Columbia*. At 9:16, the countdown arrival clock reached zero, and there was no signal or sign of the *Columbia*. Captain Bill Readdy, our Associate Administrator for Space Flight and a former astronaut, declared a space flight contingency and activated the recovery control center at the Kennedy Space Center. At that point, Bill Readdy and I departed the landing strip and headed to the launch control center.

We arrived at the launch control center 13 minutes later. At 9:29 a.m., we activated the contingency action plan for space flight operations. Through the White House situation room, we notified the President as well as other senior staff of the loss of communications. In addition, Members of Congress and the Government of Israel were notified. Homeland Security Secretary Tom Ridge and the National Security Council were also made aware of the situation and were present there in the situation room that morning.

Secretary Ridge then began assessing the possibility that this situation was terrorism related. Shortly after, he made the determination it was highly unlikely terrorism was involved. Secretary Ridge then announced that the Federal Emergency Management Agency would be the lead federal agency for the recovery effort on the ground.

Meanwhile, the family members of the *Columbia* astronauts were escorted from the landing strip to the astronauts' crew quarters. Later that morning, at about 11:30, we met with the families at the crew quarters at Kennedy Space Center to express our condolences, offer any and all support we could give, and assure them that we would offer that support throughout this entire ordeal, and stated our commitment to find the cause of the accident, fix the problems we find, and continue the work that their loved ones had started.

Data at all the NASA sites and contractors were impounded at 10 a.m., and the headquarters action team in Washington, D.C., was activated with NASA personnel moving immediately to their duty stations. By 10:30, an hour after the contingency plan had been activated, the mishap response team convened to assess the preliminary data and focus on the location of the crew compartment through the Rescue Coordination Center at Langley Air Force Base in Virginia. The rapid response team was activated for deployment to Barksdale Air Force Base in Louisiana that day.

The process of initiating the *Columbia* Accident Investigation Board began about 10:30 a.m. on Saturday, February 1st, one hour after the contingency plan was activated. I placed a call to the NASA deputy administrator, Fred Gregory, also a former astronaut, who was at NASA headquarters in Washington. Mr. Gregory then began calling the *Columbia* Accident Investigation Board members, which are specified by position in the contingency action plan.

At 1:15 that afternoon, I made a brief televised statement expressing our national regrets for the tragic accident and informed the public about the appointment of the *Columbia* Accident Investigation Board.

The Accident Investigation Board was formally activated during the NASA Mishap Investigation Team teleconference, which occurred at 5 p.m. that afternoon, Saturday, February the 1st, less than eight hours after the event.

By 6 p.m., during a teleconference with the White House situation room, we briefed officials from the Department of Homeland Security, the Federal Emergency Management Agency, the Department of Defense, the FBI, and the Federal Aviation Administration about the current status of the accident investigation.

At 6:40 that evening, staff members of the National Transportation Safety Board departed Washington and traveled to Barksdale Air Force Base in Louisiana to assist as part of the Mishap Investigation Team, that day. They were later made available to the *Columbia* Accident Investigation Board.

On Sunday, February the 2nd, the Accident Investigation Board, chaired by retired United States Navy Admiral Hal Gehman, held its first meeting at Barksdale Air Force Base in Louisiana, less than 30 hours after the accident. We also began the practice of twice-daily briefings at headquarters in Washington and at the Johnson Space Center in Houston.

Membership of the *Columbia* Accident Investigation Board consists of persons selected for their positions in heading civil and military offices with responsibility for aerospace safety, accident investigations, and related skills. Many have been chief investigators on major accidents. And between them, board members have the experience of some 50 major investigations to draw upon. Quite simply, the people who are now on the board are some of the best in the world at what they do, and they were activated immediately. You have our assurance that this distinguished board will be able to act with genuine independence.

When the board assembled, it modified its charter to eliminate any reference to NASA directing the administration of the investigation. The framework that was contained in the contingency plan was modified and will continue to be to ensure the independence of this board. NASA accepted the changes to the charter without objection, as I will continue to do in the future, as well, for any changes they propose.

Further, the NASA Inspector General Robert Cobb is an observer on the *Columbia* Accident Investigation Board, having arrived on Monday, February the 3rd. He will help assure the independence of the board, as he reports both to the President and to the Congress under the terms of the Inspector General Act.

There are additional details about the *Columbia* Accident Investigation Board and its activities that are, I think, important to note. The board has taken over hardware and software releases of NASA so that we cannot alter anything unless the board approves. We've already begun to honor document requests from the board, as we have all along, and have also supplied additional documents to the board which were not requested, but we believe might be helpful in their work as we move along. And, finally, the board has instructed NASA to conduct a fault-tree analysis that it intends to independently validate, to look at all the possible causes that could have occurred and to examine those in a very methodical way, which they will then, in turn, independently validate.

On Sunday, the NASA Mishap Investigation Team was on the ground and working with local officials in Texas and Louisiana. The State of Texas activated 800 members of the Texas National Guard to assist with the retrieval of debris, and I am eternally grateful to Governor Rick Perry for his immediate response within hours of our request.

By Tuesday, there were nearly 200 NASA and NASA contractor personnel working recovery operations in Texas, Louisiana, Arizona, and California. They were part of the more than 2,000 people from Federal Emergency Management Agency, the Environmental Protection Agency, the FBI, the Department of Defense, Department of Transportation, the U.S. Forest Service, Texas National Guard, Louisiana National Guard, and state and local authorities working to locate, document, and collect debris.

By Wednesday, the astronauts' remains were transferred to Dover Air Force Base in Delaware. At Dover, NASA Deputy Administrator Fred Gregory, and former astronaut, and ceremonial honor guard were present to pay our respect to the seven fallen astronauts.

Throughout the week, we were able to make steady progress in our efforts to recover debris from the accident. We have, thus far, recovered upwards of 12,000 elements of debris. The search effort, as you know from our press conferences, is a large, complex, and ongoing effort with hundreds of square miles with challenging weather and terrain conditions. And, indeed, the graphic that's up now is that 500-mile swath from Dallas/Fort Worth area to just south of Shreveport, Louisiana, in and around the Lufkin, Texas, area.

We're very grateful that no one was injured on the ground as a result of flying debris from the accident, and we're working with our agency partners to assure recovery operations remain safe as we continue this effort.

Throughout the course of this activity, I've also briefed the President and the Vice President on a near-daily basis to advise and apprise them of all the progress we're making, as well as the cooperation of all of the federal agencies, who have been extremely participating in this effort.

The Federal Emergency Management Agency command post was set up in Lufkin, Texas, on Saturday, the 1st of February, and has been operating nonstop since then. Debris collection activities began at Barksdale Air Force Base on Sunday, February the 2nd.

Yesterday, we began transporting debris on trucks to the Kennedy Space Center where they'll be assembled and analyzed as part of the comprehensive accident investigation directed by the Gehman board.

I visited Texas and Louisiana this past Saturday to get my own assessment of the operation, but, more importantly, to personally thank the volunteers, in addition to all the federal, state, and local public servants, who have been working so tirelessly to support the debris recovery effort.

Let me touch briefly on the Space Shuttle fleet as it is today. *Discovery* is continuing to undergo major inspections and upgrades, which will be completed by April of 2004. *Atlantis* is currently assembled and stacked in a Vehicle Assembly Building at the Kennedy Space Center for STS–114, the next mission due to have, or planned to have, been flown. The *Endeavour,* the third of the orbiters, is in the Orbiter Processing Facility and being prepared for STS–115, which was scheduled a couple of months later.

The next Shuttle mission, STS–114, was to have been to the International Space Station in March, that mission commanded by Colonel Eileen Collins, United States Air Force. And I met with her on Friday to further advise that the mission is on hold until we understand the causes of the *Columbia* accident and are able to resolve any issues identified.

At this time, we don't know how long it will be before we can resume Shuttle flights. We will only know when the *Columbia* Accident Investigation Board concludes its work and presents its findings to all of us.

Columbia was the first orbiter in the Shuttle fleet, having flown 28 successful missions, or just over a quarter of its certified life of a hundred flights. In February 2001, a little over a year ago, *Columbia* completed a major scheduled 18-month overhaul and update of its systems, a process we call "Orbiter Major Modifications." The STS–107 mission was *Columbia*'s second flight following that major overhaul. A successful servicing mission that had been conducted, the first one, was to the Hubble Space Telescope in March of 2002. So this was the second flight after it had been nearly completely rebuilt.

Prior to the loss of *Columbia* and her crew, the projected Shuttle flight rate was five per year, starting in 2004, and funding is requested for that flight rate in the budget the President just submitted last week. The flight rate will be adjusted as needed, of course, once we determine when we can return to flight safely.

The crew of the International Space Station is, of course, deeply saddened by the loss of *Columbia* and her crew, as are all of our partners and people around the world. I spoke with International Space Station crew members, Ken Bowersox, the commander, United States Navy, Don Pettit, who is our science officer aboard, and Nikolai Budarin, who is a cosmonaut engineer, on Sunday, February the 2nd for the first time in our discussions, to inform them of the accident and how we're proceeding. Despite the tragedy, the crew is continuing its busy schedule of work.

The day after the loss of STS–107, our Russian partners conducted a successful planned launch of an unmanned, autonomous Progress resupply vehicle to the station. The provisions carried on Progress 10P should provide the crew sufficient supplies to maintain normal operations through June 2003, through this summer. Progress resupply flights to the International Space Station by our Russian partners will continue as scheduled. The next flight is scheduled for June 2003.

We're working with Rosaviakosmos, the Russian Aviation and Space Agency officials, to determine what we might want to place on that flight to make sure we have the best use of the space available. In addition, a regularly scheduled Soyuz crew transport vehicle exchange is planned already for the launch in April 2003, as it had been prior to February 1st.

Study teams formed almost immediately after the accident to assess the impact on the International Space Station. These teams are focused on how we will, first, sustain the station, second, continue to assemble the station, as it is not yet complete, and, third, to maximize the utilization of this unique research platform.

We have kept our International Space Station partners informed of our recovery efforts. Further, we have met with our international partners just last week, and continue to each day, to plan future meetings in the weeks ahead to develop an International Space Station partner plan.

We can maintain a permanent crew on the International Space Station as long as it is necessary, with support from Soyuz and Progress flights. The International Space Station is stable and has sufficient propellent to maintain its orbit for at least a year without support from the Space Shuttle.

But the nearer-term issue for crew support beyond June is water. The International Space Station cannot support a crew of three after June with the currently planned support in progress. As a consequence, we're discussing with our international partners the possibility of changing the April Soyuz flight from a taxi mission to a crew exchange mission, as well as the feasibility of adding Progress resupply flights. But I want to really emphasize that there are no decisions that have been made, and all options are being examined at present.

I talked to the Expedition 6 crew that Captain Bowersix commands, now in orbit, and they've expressed determination and desire to do whatever is necessary to continue the research and deal with any changes in crew rotation schedule that may be necessary.

As we look forward to determine our nation's best course of action in responding to the *Columbia* accident, I'd like to point out that NASA developed an Integrated Space Transportation Plan, which was submitted by the President to the Congress in November as an amendment to the fiscal year 2003 budget. So three months ago, that plan was presented at that time. The Integrated Space Transportation Plan could help us address many of the near-term issues we're facing, even though it was developed prior to the loss of *Columbia*.

The plan reflects the tight coupling required across the Space Station, Space Shuttle, and the Space Launch Initiatives. It is intended to ensure that necessary access to the International Space Station can be supported for the foreseeable future. It consists of three major program elements—the Space Shuttle, the Orbital Space Plane, and the Next-Generation Launch Technology.

This new plan makes investments to extend Shuttle's operational life for continued safe operations. The Orbital Space Plane is designed to provide a crew transfer capability as early as possible to assure access to and from the International Space Station. And the Next-Generation Launch Technology program funds next-generation Reusable Launch Vehicle technology in areas such as propulsion, structures, and operation. This initiative will focus on the Orbital Space Plane and the Next-Generation Launch Technology, including third-generation Reusable Launch Vehicle efforts.

Now, the 2003 budget amendment that the President submitted last November, in 2002, also proposed adding funds to International Space Station reserves to assure that we could successfully reach the milestone of U.S. core configuration and maintain progress on the long lead items for enhanced research aboard space station and the continued buildout of that remarkable research laboratory platform.

Space flight is a means to an end at NASA. That end is research, exploration, discovery, and inspiration. The crew of STS–107 were engaged in a wide array of scientific research that could be conducted nowhere else but in space and had significant potential benefits for the public. *Columbia*'s crew took great pride in their research aimed at fighting cancer, improving crop yields, developing fire-suppression techniques, building earthquake-resistant buildings, and understanding the effects of dust storms on weather. As was recorded by the media, *Columbia* had a cargo of human ingenuity.

The crew of International Space Station is also conducting research now that cannot be conducted anywhere else. Thus far, more than 60 experiments spanning such scientific disciplines as human physiology, genetics, plant biology, Earth observations, physics, and cell biology have been conducted on the International Space Station. From these experiments, scientists are learning better methods of drug testing and about dynamic models of human diseases, the physics of fundamental processes in manufacturing, antibiotic synthesis, and changes in Earth climate, vegetation and crops.

The International Space Station is the centerpiece initiative of human space flight at NASA. Our objectives in this regard are very clear. First, we will keep on-orbit International Space Stations crews safe. Second, we intend to keep the International Space Station continuously occupied in order to assure the reliability of the station itself. And, third, we intend to return to assembly—as soon as we're able, to return the Shuttle fleet to safe operations and complete the research goals for ourselves and for our international partners.

To accomplish these aims, we need to create a long-term crew-return capability to complement and augment the Soyuz vehicles now provided by our partners. We intend to build that new return capability to create a new crew-transfer system that will allow us to rotate crews on the International Space Station independent from the Space Shuttle.

We also firmly believe that extending the operational life of the remaining Shuttle fleet is a good investment, because it will help maximize the science return from the International Space Station.

We designed our Integrated Space Transportation Plan to ensure that we have coordinated resources to exploit the unique research environment of space and the International Space Station in the near-, mid-, and long-term. We thought the plan was a good one when we proposed it, and we believe that it's not only valid today, but even more compelling to pursue. While we believe that this plan is a good one, we will re-examine it as necessary in light of the investigative findings of *Columbia*.

Just over a week ago, although it seemed more like a lifetime, the President spoke so eloquently and powerfully at the Johnson Space Center memorial service in Houston, Texas. He said, "The cause of exploration and discovery is not an option we choose; it is a desire written in the human heart. We're all part of a creation which seeks to understand all creation. We find the best among us, send them forth into unmapped darkness, and pray they will return. They go in peace for all mankind, and all mankind is in their debt."

The noble purposes described in the President's words frame all that we do and how we do it. These purposes drive our mission goals, which are to understand and protect our home planet, to explore the universe and search for life, and to inspire the next generation of explorers as only NASA can.

And even while our nonstop work to recover from this terrible tragedy and to continue safe operations on the International Space Station will be our chief focus in the days and weeks and months ahead, the American people should know we will also press ahead with other activities to achieve these important goals.

This centennial flight year, we are launching the Mars exploration rovers, the Mars spacecraft, the space infrared telescope facility, and a number of Earth science spacecraft and instruments, as well as continuing our work to help improve aviation security on behalf of our homeland defense. In these activities and in all that we do at NASA, we strive for unmatched excellence. When it comes to human space exploration, those margins are razor thin, and we know we're graded on an extremely harsh curve. For us, 96 percent to 99 percent is not an "A." One-hundred percent is the minimum passing garde.

Now, despite this harsh truth, we know the lesson from this terrible accident is not to turn our backs on exploration because it is hard or risky. John Shedd once said about the age of ocean exploration, "A ship in safe harbor is safe, but that is not what ships are built for."

Human history teaches us that in exploration, after accidents like this occur, we learn from them and further reduce risks, although we must honestly admit that risk can never be eliminated.

President John F. Kennedy observed once, some 41 years ago, speaking of our fledgling space program at that time, "All great and honorable actions are accompanied with great difficulties, and both must be enterprised and overcome with answerable courage."

The immediate task before our agency is clear. We'll find the problem that caused the loss of *Columbia* and its crew, we'll fix it, and then we'll return to flight operations that are as safe as humanly possible in pursuit of knowledge. We have no preconceptions about what caused the failure or what it will take to make it so that it will never happen again. We have an independent Accident Investigation Board of truly outstanding and eminently quality individuals, and they, and only they alone, will determine the cause of the accident and its remedy, no matter where it leads. We're ready and willing to support the addition of any experts that Admiral Gehman deems necessary to the effective conduct of the board's investigations.

Part of my job as Administrator is to remind folks of what NASA does and what we are capable of doing. It's a responsibility I take very, very seriously. And, at the same time, I am saddened beyond words at the loss of seven outstanding men and women of STS–

107. I'm also very proud and humbled by the focus, dedication, and professionalism of the NASA family and all those throughout the country who are assisting in this challenging recovery effort.

Today, February the 12th, is also the birthday of President Lincoln. And some of his words, spoken for an entirely different purpose, have come to mind this past week. "It is rather for us to be here dedicated to the great task remaining before us, that from these honored dead we take increase devotion to that cause for which they gave the last full measure of devotion."

We have an opportunity here and now to learn from this loss and renew the boundless spirit of exploration present at NASA's beginning. We will do this by being accountable to the American people for our failings and, we hope, credible and compelling in pursuit of research, exploration, and inspiration for future generations.

And, finally, during the 16-day STS–107 mission, we had no indication that would suggest a compromise to flight safety. The time it has taken me to present this testimony is about the same amount of time that transpired between when mission control first noticed anomalies in temperature measurements and the accident.

(Pause.)

I just paused for a few seconds. That's the same amount of time that transpired from mission control's last communication with the crew and our loss of signal with the heroic *Columbia* astronauts.

May Good bless the crew of STS–107.

Chairman McCain, Chairman Boehlert, thank you all very much for you attention.

[The prepared statement of Mr. O'Keefe follows:]

PREPARED STATEMENT OF SEAN O'KEEFE

Space Shuttle Columbia – STS-107

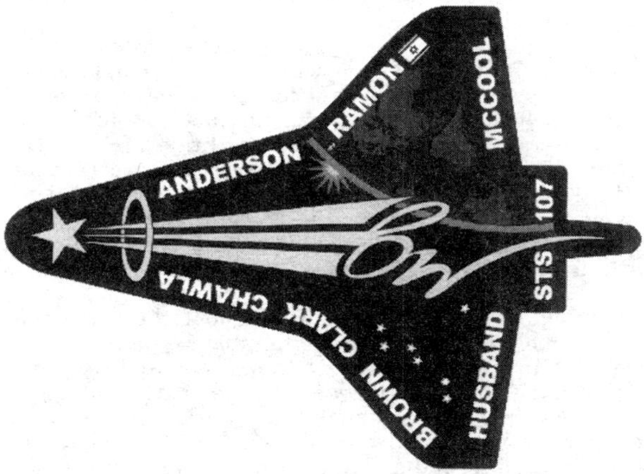

Good morning. I appreciate the opportunity to appear before this hearing of the Senate Commerce, Science and Transportation Committee and the House Science Subcommittee on Space and Aeronautics to discuss the tragic loss of the courageous crew of the Space Shuttle *Columbia,* the ongoing investigation into this tragedy, and the implications of the loss of *Columbia* to the Nation's space exploration efforts.

This morning, eleven days after the accident, our work continues to honor the solemn pledge we've made to the astronauts' families and to the American people that we will find out what caused the loss of the *Columbia* and its crew, correct what problems we find, and do our utmost to make sure this never happens again.

We welcome the Committee's interest in working with NASA to help determine how we can learn from this tragic accident so that we may continue advancing the Nation's research and exploration objectives in space while at the same time striving to ensure we make manned spaceflight as safe as humanly possible.

Throughout NASA's forty-five years of serving the public interest, Congress has been our partner, helping us achieve the goals outlined in NASA's congressionally authorized charter. This charter compels NASA to:

- Explore, use, and enable the development of space for human enterprise.
- Advance scientific knowledge and understanding of the Earth, the Solar System, and the Universe and use the environment of space for research.
- Research, develop, verify, and transfer advanced aeronautics, space, and related technologies.

With the support of Congress, NASA has amassed a record of significant achievements that have tangibly improved the lives of all Americans. When we have erred, you have helped us right our course.

This morning you will be asking us tough questions. That's as it should be. Believe me, none of the questions you will ask can be any tougher than those we are asking of ourselves.

I can assure you, however, that whatever determinations are reached regarding the cause of the accident, you will find that complacency is not one of them. Last year we temporarily halted Shuttle flight operations when tiny cracks were discovered in metal liners used to direct the fuel flow inside propellant lines on two different orbiters. We did not fly again until that problem was corrected. To signal our continued commitment to rewarding such diligence, we also made a point to praise David Strait, the young contractor employee who discovered the cracks. Other flight decisions made throughout the year were made with the goal of operational safety being paramount. And from working with the dedicated employees who keep the Shuttle flying safely I know they have the utmost regard for the enormity of their duty.

This week, at NASA Centers throughout the country and in the field, with the support of more than 2000 people from more than 20 federal, state and local organizations, the important work of data analysis and recovery operations is continuing. I am extremely proud of the work that is being conducted by these dedicated public servants. As President Bush said last week, "The people of NASA are being tested once again. In your grief, you are responding as your friends would have wished—with focus, professionalism, and unbroken faith in the mission of this agency. Captain Dave Brown was correct: America's space program will go on."

This morning, to help frame our discussion, I would like to review for you the significant actions NASA has taken since the morning of the accident in accord with our contingency plan. In addition to articulating notification or first response procedures, defining the roles and responsibilities of mishap response and mishap investigation teams, the plan specifies selection of persons outside of NASA to head an independent, seasoned accident investigation team.

While we did not foresee this terrible tragedy, our response has unfolded as we had planned and prepared for that contingency plan. This plan was one of the many positive outcomes from the terrible loss of the Space Shuttle *Challenger* seventeen years ago. The plan is updated before every flight and a contingency was simulated just three months ago.

First Response: Saturday February 1, 2003

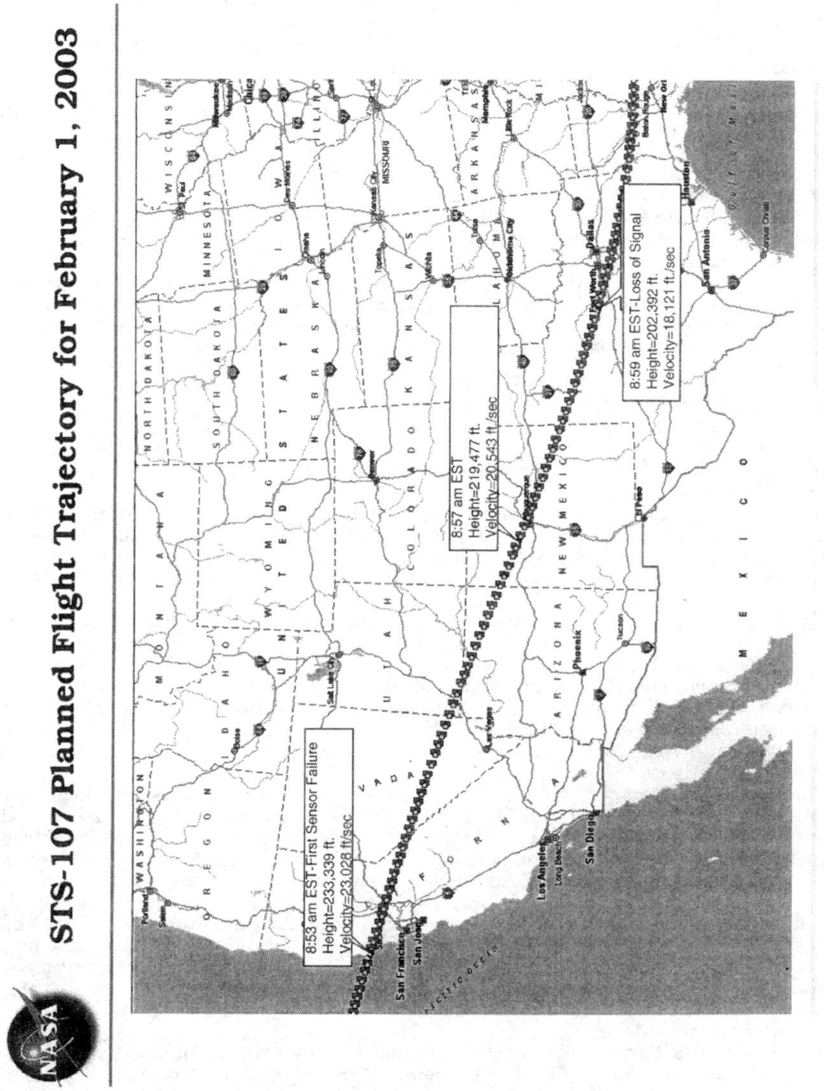

When we first became aware of the a problem with STS–107, I was waiting at the Space Shuttle Landing Strip at the Kennedy Space Center on Saturday, February 1. At 8:59 a.m. eastern time, we lost communications with the *Columbia.*

At 9:16 a.m. the countdown arrival clock reached zero and there was no sign of the *Columbia.* Captain Bill Readdy, our Associate Administrator for Space Flight, declared a spaceflight contingency and activated the Recovery Control Center at the Kennedy Space Center. At that point, Bill Readdy and I departed the landing strip and headed to the Launch Control Center.

We arrived at the Launch Control Center thirteen minutes later, at 9:29 a.m., and we activated the Contingency Action Plan for Space Flight Operations. Through the White House Situation Room, we notified the President as well as other senior staff of the loss of communication. In addition, Members of Congress and the Government of Israel were notified. Homeland Security Secretary Tom Ridge and the National Security Council were also made aware of the situation. Secretary Ridge then began assessing the possibility that this situation was terrorism-related. Shortly after, he made a determination that it was highly unlikely terrorism was involved.

Secretary Ridge then announced that the Federal Emergency Management Agency would be the lead federal agency for the recovery effort.

Meanwhile, the family members of the *Columbia* astronauts were escorted from the landing strip to the astronauts' crew quarters. Later that morning, at about 11:30 a.m., I met with the families at the crew quarters at Kennedy Space Center to express my condolences, offering any and all support we could give, and stated our commitment to find the cause of the accident, fix any problems we may find, and continue the work that their loved ones had started.

Data at all NASA sites and contractors were impounded at 10:00 a.m. and the Headquarters Action Center in Washington, D.C. was activated with NASA personnel moving immediately to their duty stations.

By 10:30 a.m., the NASA Mishap Response Team convened to assess the preliminary data and focus on the location of the crew compartment through the Rescue Coordination Center at Langley Air Force Base in Virginia. The Rapid Response Team was activated for deployment to Barksdale AFB in Louisiana.

Columbia Accident Investigation Board

The process of initiating the *Columbia* Accident Investigation Board began about 10:30 a.m. on Saturday, February 1, when I placed a call to NASA Deputy Administrator Fred Gregory, who was at NASA Headquarters in Washington. Mr. Gregory then began calling *Columbia* Accident Investigation Board members currently listed in our contingency plan.

At 1:15 p.m., I made a brief televised statement expressing our "deepest national regrets" for the tragic accident and informed the public about the appointment of the *Columbia* Accident Investigation Board.

I verbally activated the *Columbia* Accident Investigation Board during the NASA Mishap Investigation Team teleconference, which occurred at 5:00 p.m.

By 6:00 p.m. during a teleconference with the White House Situation Room, we briefed officials from the Department of Homeland Security, the Federal Emergency Management Agency, the Department of Defense, the FBI, and the Federal Aviation Administration about the current status of the accident investigation.

At 6:40 p.m. staff members of the National Transportation Safety Board departed Washington and traveled to Barksdale Air Force Base in Louisiana to assist as part of the Mishap Investigation Team. They were later made available to the *Columbia* Accident Investigation Board.

On Sunday, February 2, the *Columbia* Accident Investigation Board, headed by retired U.S. Navy Admiral Hal Gehman, held its first meeting at Barksdale AFB, less than 30 hours after the accident. We also began the practice of twice daily briefings at Headquarters in Washington and at the Johnson Space Center in Houston.

Membership in the *Columbia* Accident Investigation Board consists of persons selected for their positions in heading civil and military offices with responsibilities for aerospace safety accident investigations and related skills. Many have been chief investigators on major accidents and between them the *Columbia* Accident Investigation Board members have the experience of some 50 major investigations to draw upon.

Quite simply, the people who are now on the Board are some of the best in the world at what they do.

You have our assurance that this distinguished Board will be able to act with genuine independence. When the Board assembled, it modified its Charter to eliminate any reference to NASA directing the administration of the investigation. NASA accepted the changes to the Charter without objection. Further, the NASA Inspector General, Robert Cobb is an observer on the *Columbia* Accident Investigation Board and he will help assure the independence of the Board as he reports to the President and Congress.

There are some additional details about the *Columbia* Accident Investigation Board and its activities that are worth noting. The Board has taken over hardware and software releases of NASA so that NASA cannot alter anything unless the Board approves. NASA has already begun to honor document requests from the Board, and has also supplied additional documents to the Board which were not requested that we believe may be helpful to their work. And finally,. the Board has instructed NASA to conduct fault tree analysis that it intends to independently validate.

Recovery Operations

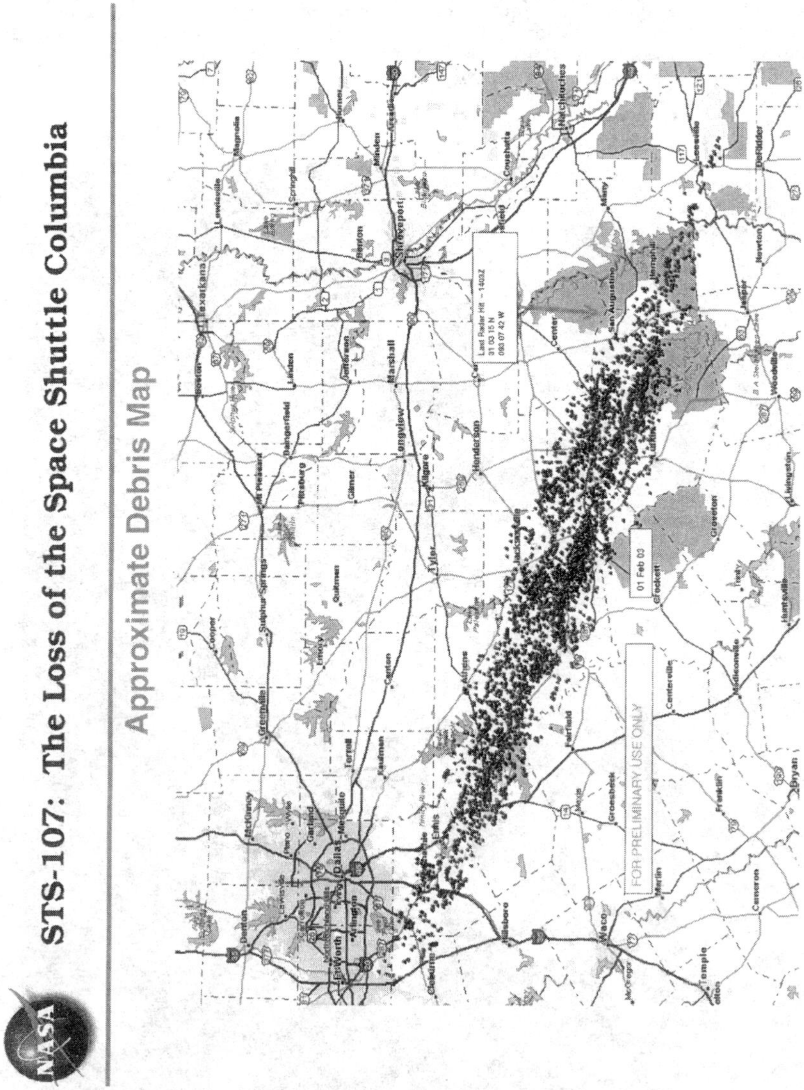

On Sunday, the NASA Mishap Investigation Team was on the ground and working with local officials in Texas and Louisiana. The State of Texas activated 800 members of the Texas National Guard to assist with the retrieval of debris.

By Tuesday, there were nearly 200 NASA and NASA contractor personnel working recovery operations in Texas, Louisiana, Arizona, and California. They were part of the more than 2000 people from Federal Emergency Management Agency, Environmental Protection Agency, Federal Bureau of Investigation, Department of Defense, Department of Transportation, U.S. Forest Service, Texas National Guard, and state and local authorities working to locate, document, and collect debris.

By Wednesday, the astronauts' remains were transported to Dover Air Force Base in Delaware. At Dover, NASA Deputy Administrator Fred Gregory and a ceremonial honor guard were present to pay our respects to the seven fallen astronauts.

Throughout the week, we were able to make steady progress in our effort to recover debris from the accident. We have thus far recovered upwards of 12,000 elements of debris. The search effort, as you know from our press conferences, is a large, complex and ongoing effort over hundreds of square miles with challenging weather and terrain conditions. We are very grateful that no one was injured on the ground as a result of flying debris from the accident and we are working with our agency partners to ensure recovery operations remain safe.

The Federal Emergency Management Agency command post was set up in Lufkin, Texas on Saturday, February 1, and has been operating non-stop since then. Debris collection activities began at Barksdale Air Force Base on Sunday, February 2. Yesterday, we began transporting debris on trucks to the Kennedy Space Center where they will be assembled and analyzed as part of the comprehensive accident investigation directed by the Gehman Board. I visited Texas and Louisiana this past Saturday to get my own assessment of the operation and to personally thank the many volunteers who have worked so tirelessly to support the debris recovery effort.

Space Shuttle Status

Let me touch briefly on the Space Shuttle fleet as it is today. *Discovery* is continuing to undergo major inspections and upgrades which will be completed by April of 2004. *Atlantis* is currently assembled and stacked in the Vehicle Assembly Building at the Kennedy Space Center for STS–114. The *Endeavour* is in the Orbiter Processing Facility and being prepared for STS–115.

The next Shuttle mission, STS–114, was to have been to the International Space Station in March. That mission, commanded by Col. Eileen Collins, U.S. Air Force, is on hold until we understand the causes of the *Columbia* accident and are able to resolve any issues identified. At this time we don't know how long it will be before we can resume Shuttle flights. We will only know when the *Columbia* Accident Investigation Board concludes its work and presents its findings.

Columbia was the first Orbiter in the Shuttle fleet, having flown 28 successful missions or just over a quarter of its certified life of 100 flights. In February 2001, less than a year ago, *Columbia* completed a major scheduled eighteen month overhaul and update of its systems, a process we call Orbiter Major Modifications (OMM).

The STS–107 mission was *Columbia's* second flight following OMM and a successful servicing mission to the Rubble Space Telescope in March 2002.

Prior to the loss of *Columbia* and her crew, the projected Shuttle flight rate was five flights per year starting in FY 2004, and we have requested funding for that flight rate in this budget. The flight rate will be adjusted as needed once we determine when we can return to flight.

International Space Station Status

The crew of the International Space Station is of course deeply saddened by the loss of *Columbia* and her crew—as are all of our partners and people around the world. I spoke with International Space Station crew members Ken Bowersox, Don Pettit, and Nikolai Budarin on Saturday, February 1st to inform them of the accident and how we are proceeding. Despite this tragedy, the crew is continuing its busy schedule of work.

The day after the loss of STS–107, our Russian partners conducted a successful launch of an unmanned, autonomous Progress resupply vehicle to the Station. The provisions carried on Progress 10P should provide the crew sufficient supplies to maintain normal operations through June 2003.

Progress resupply flights to the International Space Station by our Russian partner will continue as scheduled. The next Progress flight is scheduled for June 8, 2003. We are working with the Russian Aviation and Space Agency officials to determine what we want to place on the flight to make sure we make the best use of the space available. In addition, a regularly scheduled Soyuz crew transport vehicle exchange is already planned for launch in April 2003.

Study teams formed almost immediately after the accident to assess the impact on the International Space Station. These teams are focused on how we will 1) sustain the Station, 2) continue to assemble the Station, and 3) maximize the utilization of this unique research platform. We have kept our International Space Station partners informed of our recovery efforts. Further, we met with our international partners last week and plan future meetings in the weeks ahead to develop an International Space Station partner plan.

We can maintain a permanent crew on the International Space Station as long as is necessary with support from Soyuz and Progress flights. The International Space Station is stable and has sufficient propellant to maintain its orbit for at least a year without support from the Space Shuttle. A nearer, term issue for crew support beyond June is water. The International Space Station cannot support a crew of three after June with the currently planned support from Progress. As a consequence, we are discussing with our international partners the possibility of changing the April Soyuz flight from a taxi mission to a crew exchange mission as well as the feasibility of adding Progress resupply flights.

I should emphasize however, that no decisions have been made and we are examining all options. I have talked to the Expedition Six crew now on-orbit and they have expressed determination and desire to do whatever is necessary to continue their research and deal with any changes in the crew rotation schedule that may be necessary.

Integrated Space Transportation Plan

As we look forward to determine our nation's best course of action in response to the *Columbia* accident, it is worth noting NASA's Integrated Space Transportation Plan (ISTP), which was submitted by the President to Congress in November as an amendment to the Fiscal Year 2003 federal budget. The Integrated Space Transportation Plan can help us address many of the near-term issues we are facing, even though it was developed prior to the loss of *Columbia*.

The Integrated Space Transportation Plan reflects the tight coupling required across the Space Station, Space Shuttle, and Space Launch Initiatives efforts. It is intended to ensure that necessary access to the International Space Station can be supported for the foreseeable future. It consists of three major programs: the Space Shuttle, the Orbital Space Plane, and Next Generation Launch Technology.

- The new plan makes investments to *extend Shuttle's operational life* for continued safe operations.
- The Orbital Space Plane is designed to *provide a crew transfer capability as early as possible* to assure access to and from the International Space Station.
- The Next Generation Launch Technology Program *funds next generation reusable launch vehicle technology* developments in areas such as propulsion, structures, and operations.
- The SLI will focus on the Orbital Space Plane and Next Generation Launch Technology, including Third Generation RLV efforts.

The FY 2003 budget amendment also proposed adding funds to International Space Station reserves to assure that we could successfully reach the milestone of U.S. core complete and maintain progress on long-lead items for enhanced research aboard the Space Station.

Science and Research Objectives

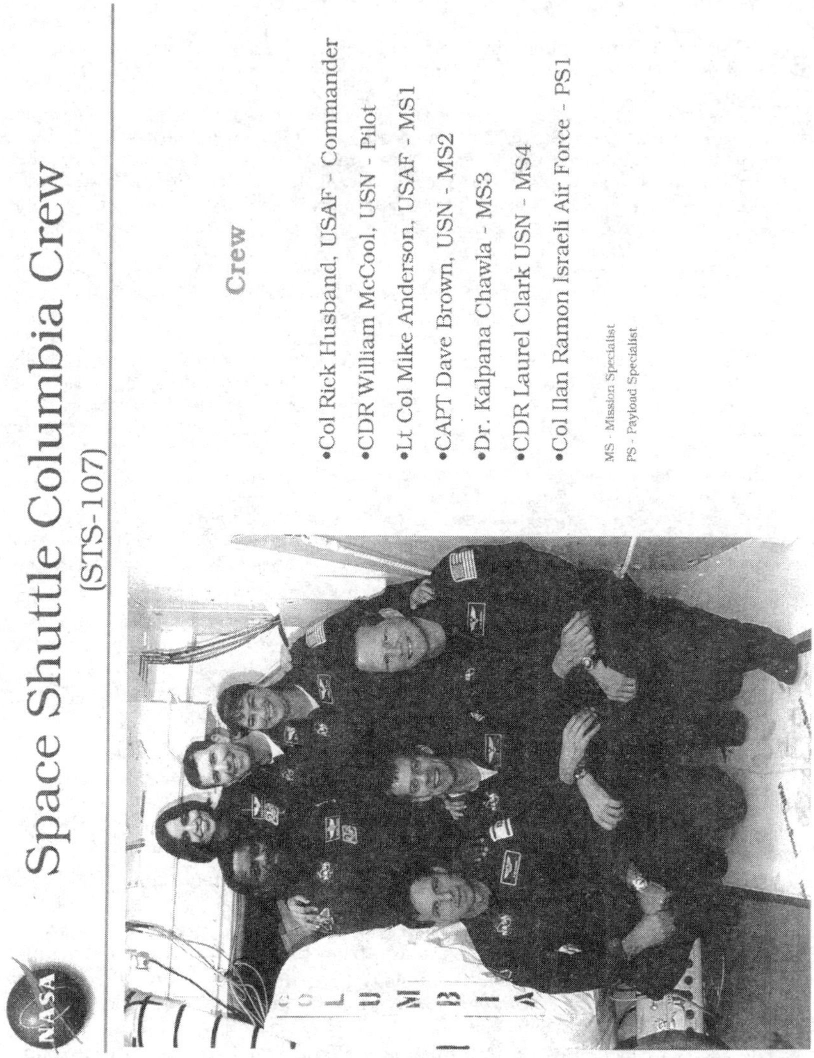

Space flight is a means to an end and at NASA that end is research, exploration, discovery and inspiration.

The crew of STS–107 were engaged in a wide array of scientific research that could be conducted nowhere else but in space, and had significant potential benefits for the public. *Columbia's* crew took great pride in their research aimed at fighting cancer, improving crop yields, developing fire-suppression techniques, building earthquake-resistant buildings, and understanding the effects of dust storms on weather. As was written in the press, "*Columbia* had a cargo of human ingenuity."

The crew of the International Space Station is also conducting research now that can be conducted nowhere else. Thus far, more than sixty experiments spanning across such scientific disciplines as human physiology, genetics, plant biology, Earth observations, physics, and cell biology have been conducted on the International

Space Station. From these experiments scientists are learning better methods of drug testing, and about dynamic models of human diseases, the physics of fundamental processes in manufacturing, antibiotic synthesis, and changes in Earth climate, vegetation, and crops.

The International Space Station is the centerpiece initiative of human space flight at NASA. Our objectives in this regard are very clear. First, we will keep our on-orbit International Space Station crew safe. Second, we intend to keep the International Space Station continuously occupied in order to assure the reliability of the International Space Station itself. Third, we intend to return to assembly as soon as we are able to return the Shuttle fleet to safe operations, and complete the research goals set for ourselves and our international partners.

To accomplish these aims, we need to create a long-term crew return capability to complement and augment the Soyuz vehicles now provided by our Russian partners. We intend to build on that new return capability to create a crew transfer system that will allow us to rotate crews on the International Space Station independently from the Space Shuttle.

We also firmly believe that extending the operational life of the remaining Shuttle fleet is a good investment because it will help maximize the science return from the International Space Station.

We designed our Integrated Space Transportation Plan (ISTP) to ensure that we had the coordinated resources to exploit the unique research environment of space and the International Space Station in the near-, mid-, and long-term.

We thought the plan was a good one when we proposed it and we believe that it is not only valid today but even more compelling to pursue. While we believe the ISTP is a good plan, we will re-examine it if necessary in light of investigation findings on *Columbia*.

Moving Forward

Just over a week ago—although it seems more like a lifetime—the President spoke eloquently and powerfully at the Johnson Space Center in Houston, Texas. He said:

> "The cause of exploration and discovery is not an option we choose; it is a desire written in the human heart. We are that part of creation which seeks to understand all creation. We find the best among us, send them forth into unmapped darkness, and pray they will return. They go in peace for all mankind, and all mankind is in their debt."

The noble purposes described in President Bush's words frames all that we do and how we do it. These purposes drive our mission goals, which are:

> To understand and protect our home planet; To explore the Universe and search for life; and, To inspire the next generation of explorers as only NASA can.

And even while our nonstop work to recover from this terrible tragedy and to continue safe operations on the International Space Station will be our chief focus in the days ahead, the American people should know that we will also press ahead with our other activities to achieve these important goals.

This Centennial of Flight year we will be launching the Mars Exploration Rovers, the Mars Express spacecraft, the Space InfraRed Telescope Facility, and a number of Earth Science spacecraft and instruments, as well as continuing our work to help improve aviation security on behalf of our Homeland Defense.

In these activities and in all we do at NASA, we strive for unmatched excellence. And when it comes to human space exploration, where margins are razor thin, we know we are graded on a very harsh curve. For us, ninety-six percent to ninety-nine percent is not an "A." One hundred percent is the minimum passing grade.

Despite this harsh truth, we know the lesson from this terrible accident is not to turn our backs on exploration simply because it is hard or risky. As John Shedd wrote about the age of ocean exploration, "A ship in harbor is safe, but that is not what ships are built for." Human history teaches us that in exploration, after accidents like this occur, we can learn from them and further reduce risk, although we must honestly admit that risks can never be eliminated. And as President John F. Kennedy said some 41 years ago, speaking about our fledgling space program, "All great and honorable actions are accompanied with great difficulties, and both must be enterprised and overcome with answerable courage."

The immediate task before the Agency is clear. We will find the problem that caused the loss of *Columbia* and its crew, we will fix it, and we will return to flight operations that are as safe as humanly possible in pursuit of knowledge. We have no preconceptions about what the cause of failure was or what it will take to make sure it never happens again. We have an independent accident investigation board of truly outstanding and eminently qualified individuals and they, and they alone, will determine the cause of the accident and its remedy—no matter where it leads.

We are ready and willing to support the addition of any experts that Admiral Gehman deems necessary to the effective conduct of the Board's investigations.

Part of my job as Administrator is to remind everyone of what NASA does and what we are capable of doing. It's a responsibility I take very seriously. At the same time that I am saddened beyond words for the loss of the seven outstanding men and women of STS–107, I am also very proud and humbled by the focus, dedication and professionalism of the NASA family and all those throughout the country who are assisting us in the recovery effort.

Today, February 12, is also the birthday of President Lincoln. And some of his words, spoken for a very different purpose, have come to be in my mind this past week:

> "It is rather for us to be here dedicated to the great task remaining before us—that from these honored dead we take increased devotion to that cause for which they gave the last full measure of devotion."

We have an opportunity here and now to learn from this loss, and renew the boundless spirit of exploration present at NASA's beginning. We will do this by being accountable to the American people for our failings and, we hope, credible and compelling in pursuit of research, exploration, and inspiration for future generations.

Finally, during the 16-day STS–107 mission we had no indications that would suggest a compromise to flight safety. The time it took me to present this testimony is about the same amount of time that transpired between when Mission Control first noticed anomalies in temperature measurements and the accident.

I just paused for a few seconds. That's the same amount of time that transpired from Mission Control's last communication with the crew and our loss of signal with the heroic *Columbia* astronauts.

May God bless the crew of STS–107.

Backup

Official STS-107 Crew Picture

Crew

- Col Rick Husband, USAF - Commander
- CDR William McCool, USN - Pilot
- Lt Col Mike Anderson, USAF - MS1
- CAPT Dave Brown, USN - MS2
- Dr. Kalpana Chawla - MS3
- CDR Laurel Clark USN - MS4
- Col Ilan Ramon Israeli Air Force - PS1

MS - Mission Specialist
PS - Payload Specialist

Liftoff

STS-107 Crew Doing Research

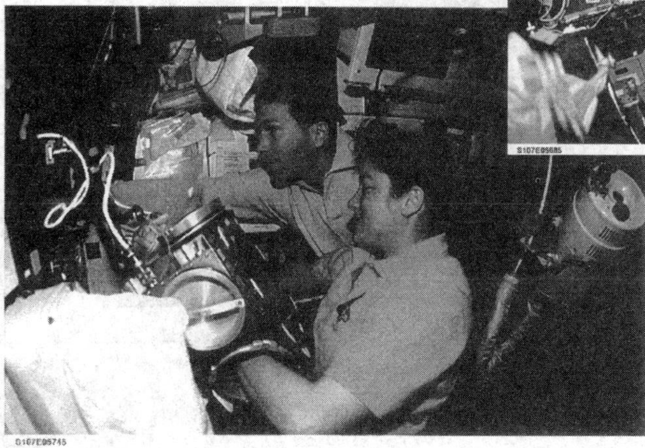

STS-107 Stack

External Tank

Debris Falling from External Tank

Cross-section of Orbiter Left Wing

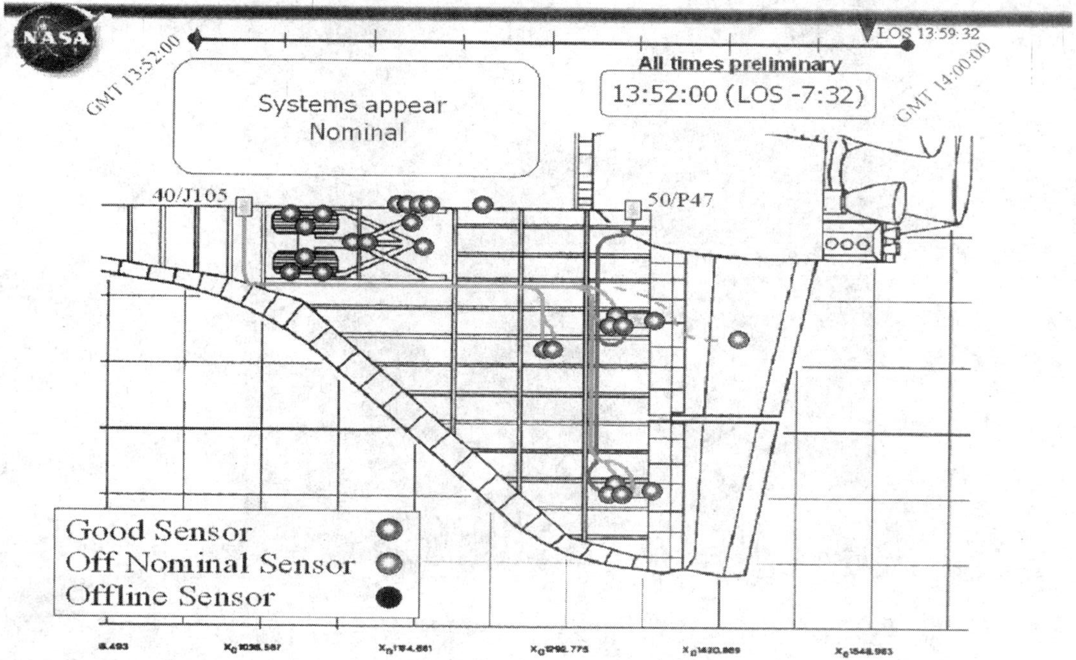

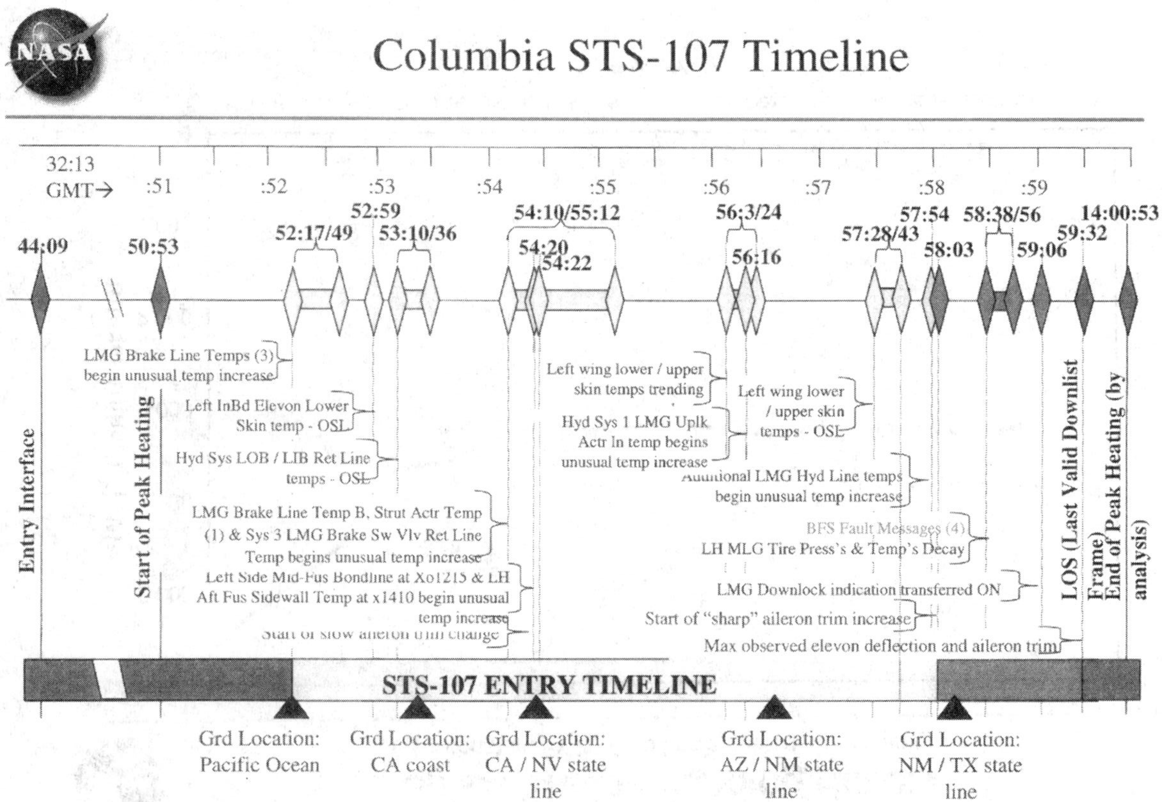
Columbia STS-107 Timeline
32:13
GMT→
:51
:52
:53
:54
:55
:56
:57
:58
:59
44:09
50:53
52:17/49
52:59
53:10/36
54:10/55:12
54:20
54:22
56:3/24
56:16
57:28/43
57:54
58:03
58:38/56
59:06
59:32
14:00:53
Entry Interface
Start of Peak Heating
LMG Brake Line Temps (3) begin unusual temp increase
Left InBd Elevon Lower Skin temp - OSL
Hyd Sys LOB / LIB Ret Line temps - OSL
LMG Brake Line Temp B, Strut Actr Temp (1) & Sys 3 LMG Brake Sw Vlv Ret Line Temp begins unusual temp increase
Left Side Mid-Fus Bondline at Xo1215 & LH Aft Fus Sidewall Temp at x1410 begin unusual temp increase
Start of slow aileron trim change
Left wing lower / upper skin temps trending
Hyd Sys 1 LMG Uplk Actr In temp begins unusual temp increase
Left wing lower / upper skin temps - OSL
Additional LMG Hyd Line temps begin unusual temp increase
BFS Fault Messages (4)
LH MLG Tire Press's & Temp's Decay
LMG Downlock indication transferred ON
Start of "sharp" aileron trim increase
Max observed elevon deflection and aileron trim
LOS (Last Valid Downlist Frame)
End of Peak Heating (by analysis)
STS-107 ENTRY TIMELINE
Grd Location: Pacific Ocean
Grd Location: CA coast
Grd Location: CA / NV state line
Grd Location: AZ / NM state line
Grd Location: NM / TX state line

Space Shuttle Funding History

(in constant FY 2003 dollars adjusted for inflation)

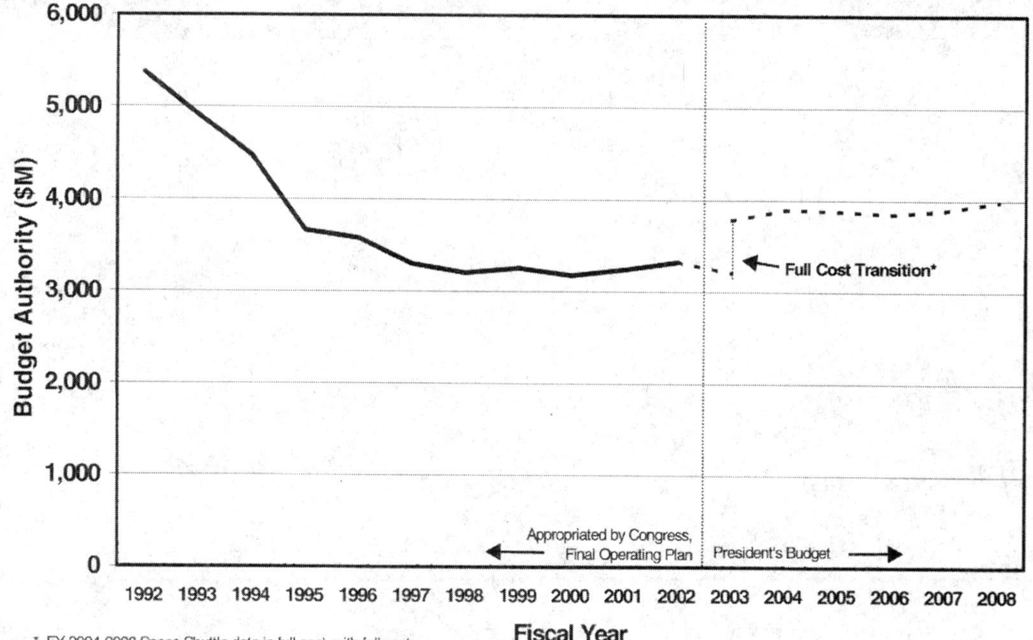

* FY 2004-2008 Space Shuttle data in full cost, with full cost estimate for FY 2003 shown to indicate transition

Space Shuttle Upgrades

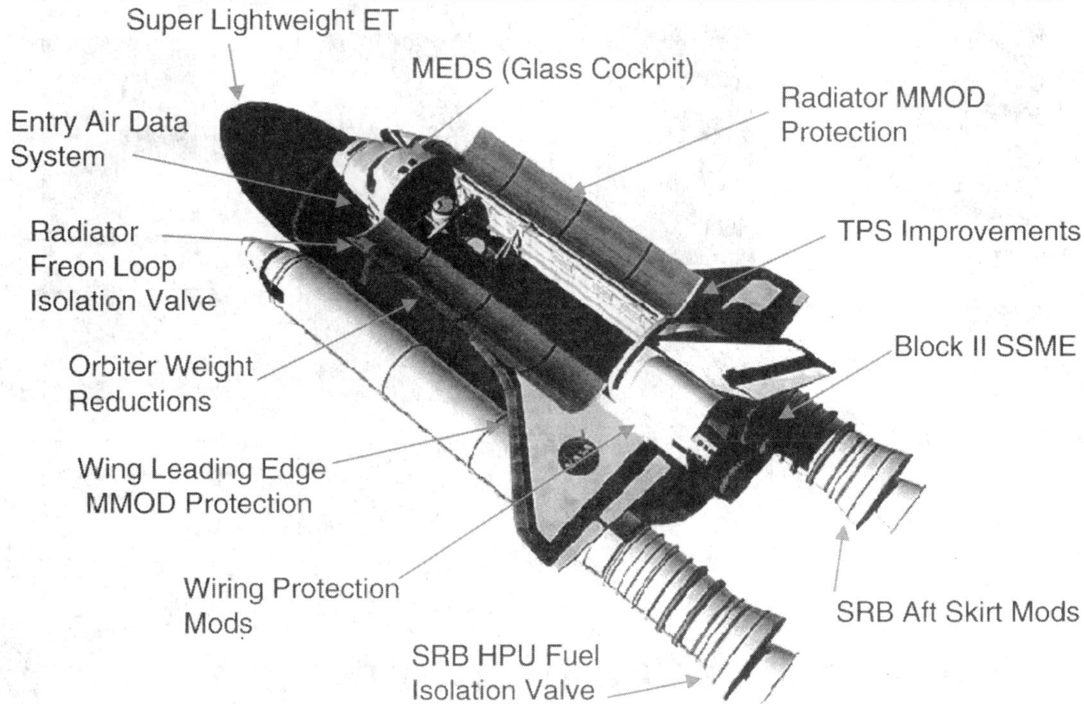

Significant Shuttle Improvements

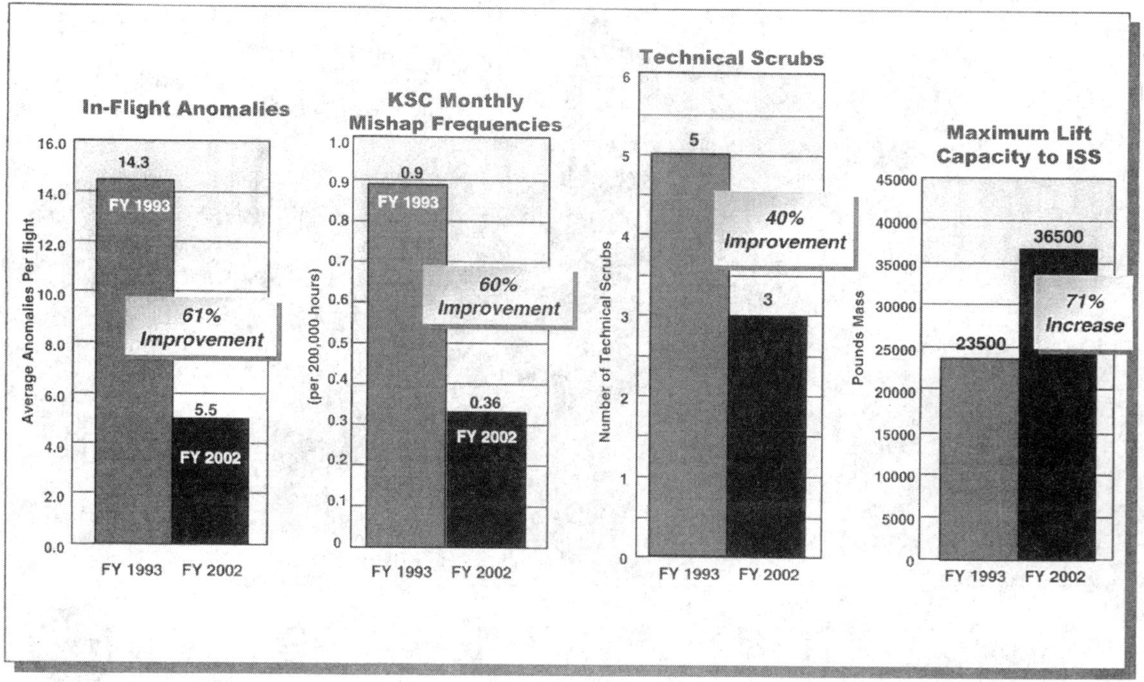

Safety Indicators

Workforce Accidents Down 26 Percent

OSHA Recordable Injury/Illness Rate

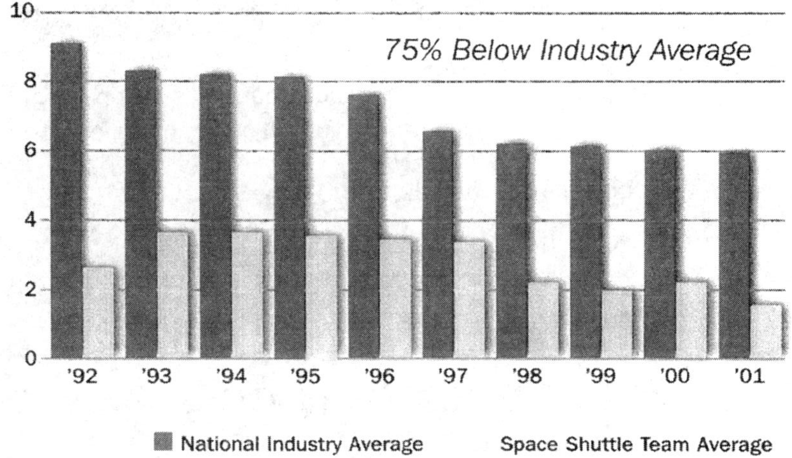

Safety Indicators

...Launches More Than Ever...

Cargo Capacity Up As Much As 100%

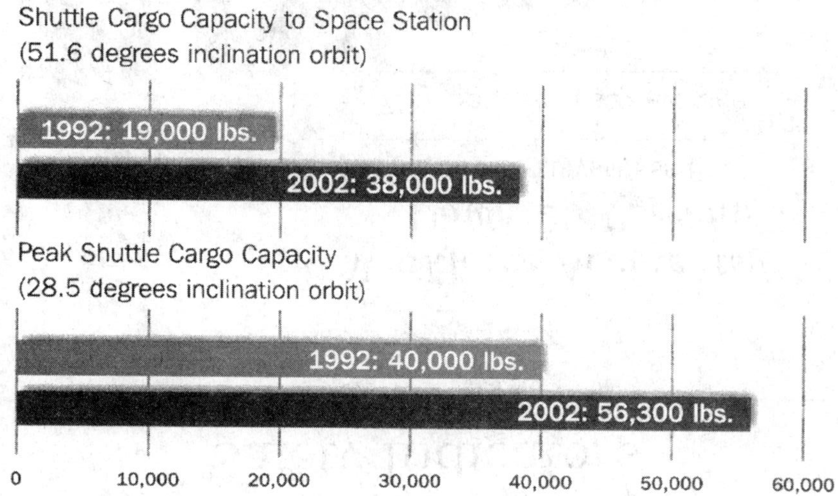

Safety Indicators

In-flight Problems
Down 70 Percent

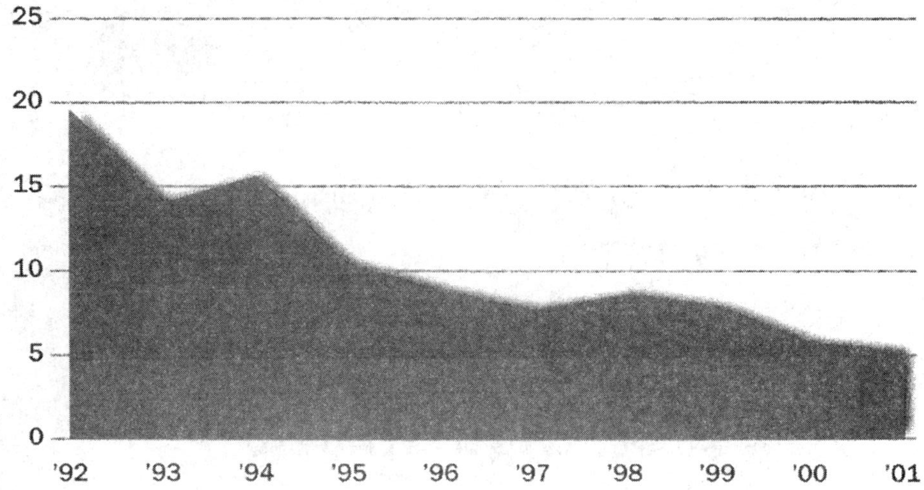

Safety Indicators

...In Less Time

Time to Plan, Design, Prepare a Shuttle Mission

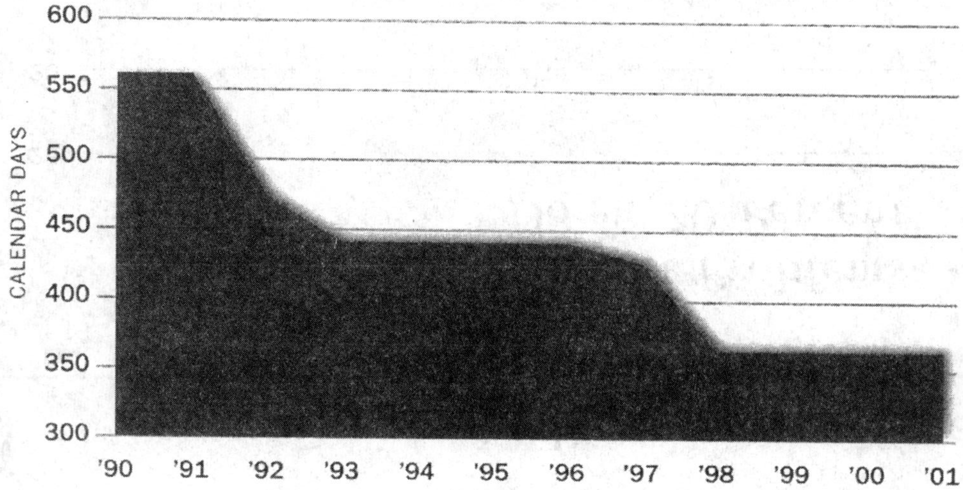

OV-104 MPS LH2 Feedline Flowliner Cracks

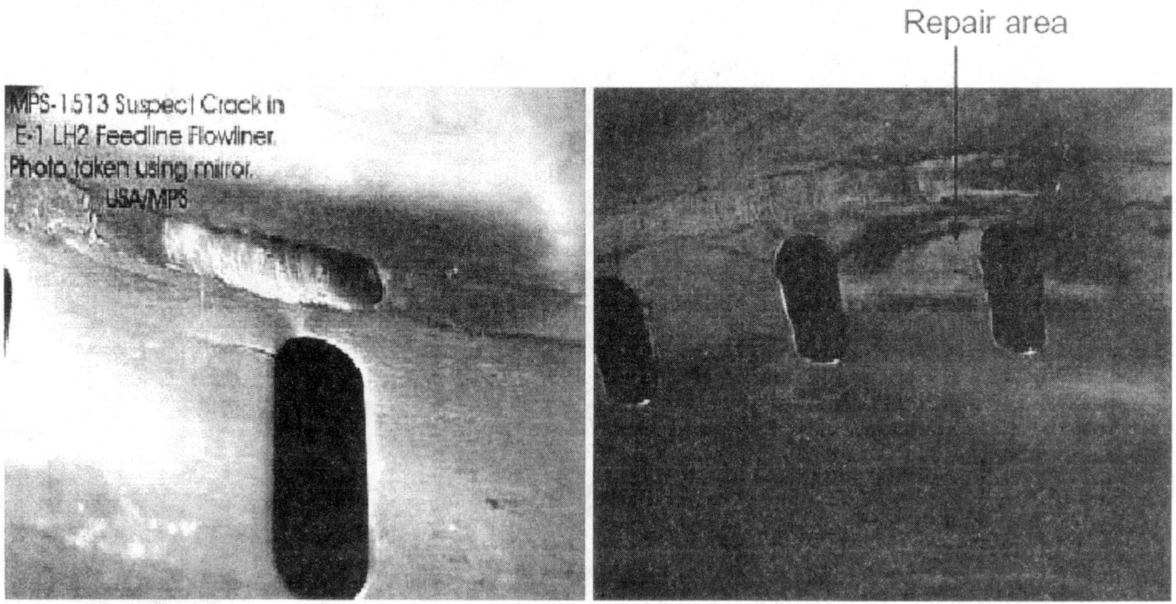

ISS – Current Configuration

International Space Station Expedition 6

Commander Ken Bowersox Works on the Zeolite Crystal Growth Experiment

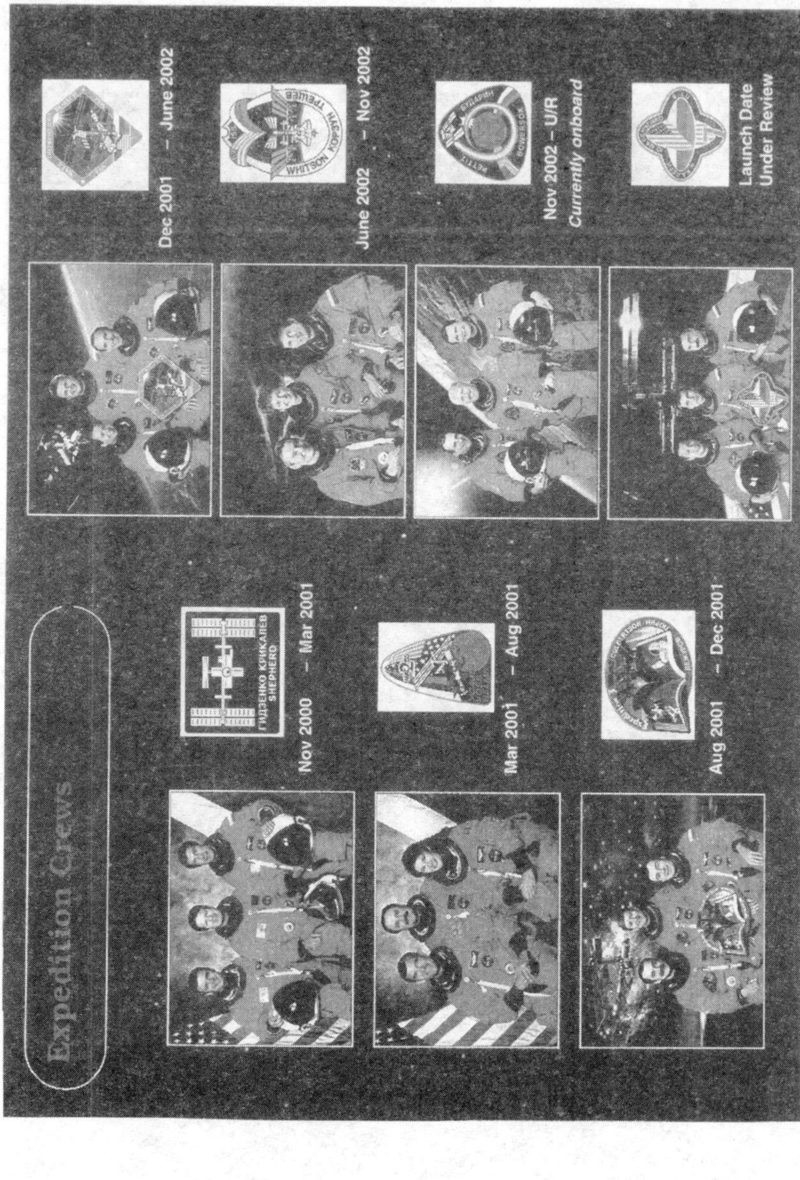
Expedition Crews
Nov 2000 – Mar 2001
ГИДЗЕНКО КРИКАЛЕВ
SHEPHERD
Mar 2001 – Aug 2001
Aug 2001 – Dec 2001
Dec 2001 – June 2002
June 2002 – Nov 2002
Nov 2002 – U/R
Currently onboard
Launch Date
Under Review

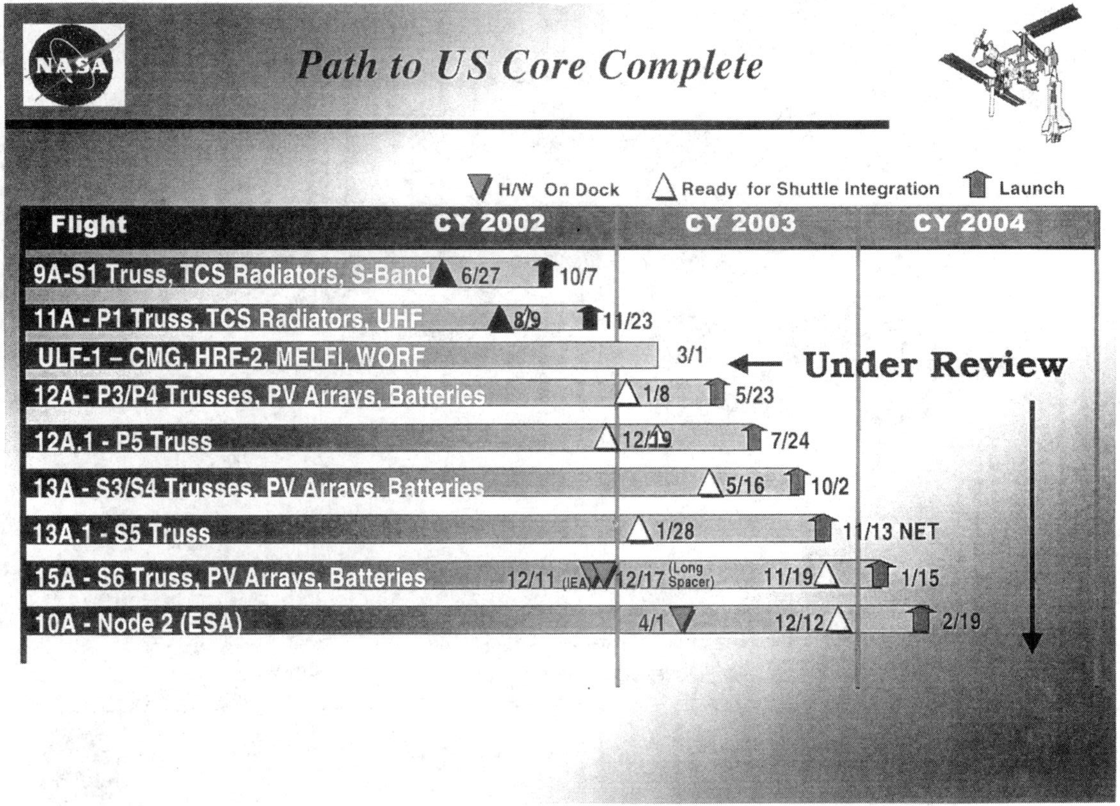
NASA
Path to US Core Complete
H/W On Dock
Ready for Shuttle Integration
Launch
Flight
CY 2002
CY 2003
CY 2004
9A-S1 Truss, TCS Radiators, S-Band
6/27
10/7
11A - P1 Truss, TCS Radiators, UHF
8/9
11/23
ULF-1 – CMG, HRF-2, MELFI, WORF
3/1
Under Review
12A - P3/P4 Trusses, PV Arrays, Batteries
1/8
5/23
12A.1 - P5 Truss
12/19
7/24
13A - S3/S4 Trusses, PV Arrays, Batteries
5/16
10/2
13A.1 - S5 Truss
1/28
11/13 NET
15A - S6 Truss, PV Arrays, Batteries
12/11 (IEA)
12/17 (Long Spacer)
11/19
1/15
10A - Node 2 (ESA)
4/1
12/12
2/19

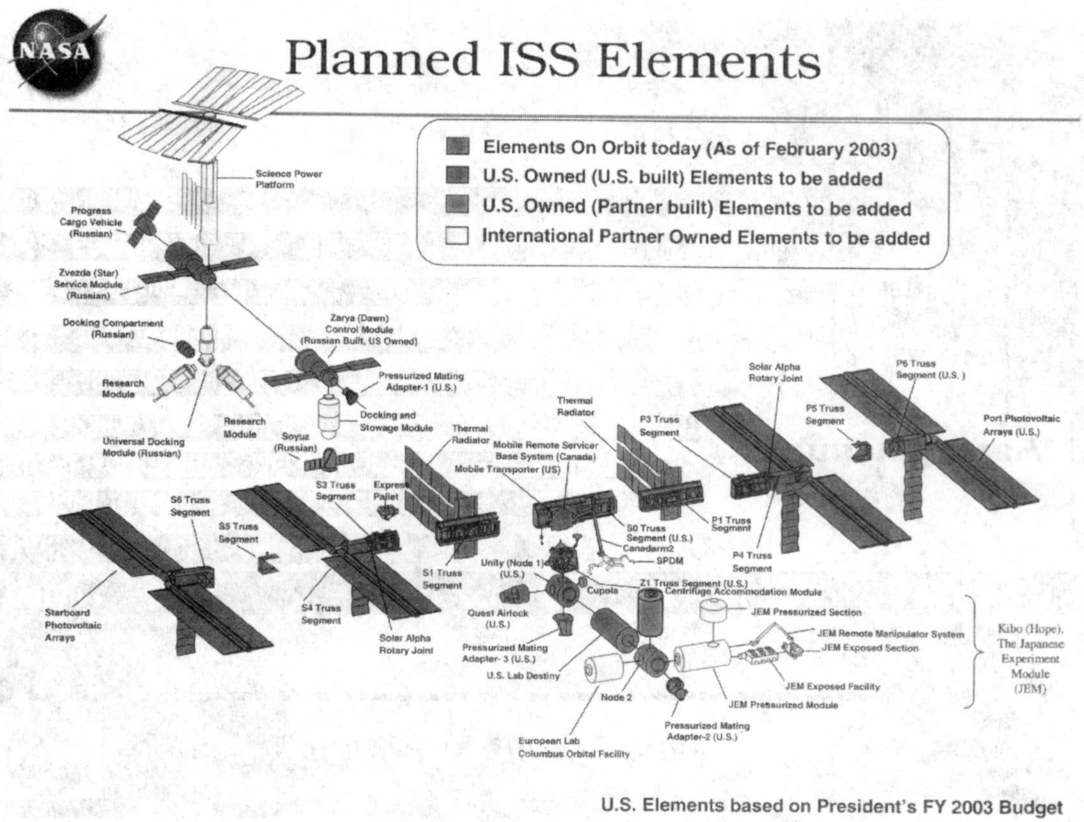
NASA
Planned ISS Elements
Elements On Orbit today (As of February 2003)
U.S. Owned (U.S. built) Elements to be added
U.S. Owned (Partner built) Elements to be added
International Partner Owned Elements to be added
Science Power Platform
Progress Cargo Vehicle (Russian)
Zvezda (Star) Service Module (Russian)
Docking Compartment (Russian)
Zarya (Dawn) Control Module (Russian Built, US Owned)
Pressurized Mating Adapter-1 (U.S.)
Research Module
Research Module
Docking and Stowage Module
Universal Docking Module (Russian)
Soyuz (Russian)
Thermal Radiator
Mobile Remote Servicer Base System (Canada)
Mobile Transporter (US)
Thermal Radiator
P3 Truss Segment
Solar Alpha Rotary Joint
P5 Truss Segment
P6 Truss Segment (U.S.)
Port Photovoltaic Arrays (U.S.)
S6 Truss Segment
S5 Truss Segment
S3 Truss Segment
Express Pallet
S1 Truss Segment
S0 Truss Segment (U.S.)
Canadarm2
SPDM
P1 Truss Segment
P4 Truss Segment
Unity (Node 1) (U.S.)
Cupola
Z1 Truss Segment (U.S.)
Centrifuge Accommodation Module
Starboard Photovoltaic Arrays
S4 Truss Segment
Solar Alpha Rotary Joint
Quest Airlock (U.S.)
Pressurized Mating Adapter- 3 (U.S.)
U.S. Lab Destiny
JEM Pressurized Section
JEM Remote Manipulator System
JEM Exposed Section
JEM Exposed Facility
JEM Pressurized Module
Node 2
Pressurized Mating Adapter-2 (U.S.)
European Lab Columbus Orbital Facility
Kibo (Hope), The Japanese Experiment Module (JEM)
U.S. Elements based on President's FY 2003 Budget

Progress Vehicle

Soyuz Vehicle

CAIB Board

CAIB on Site

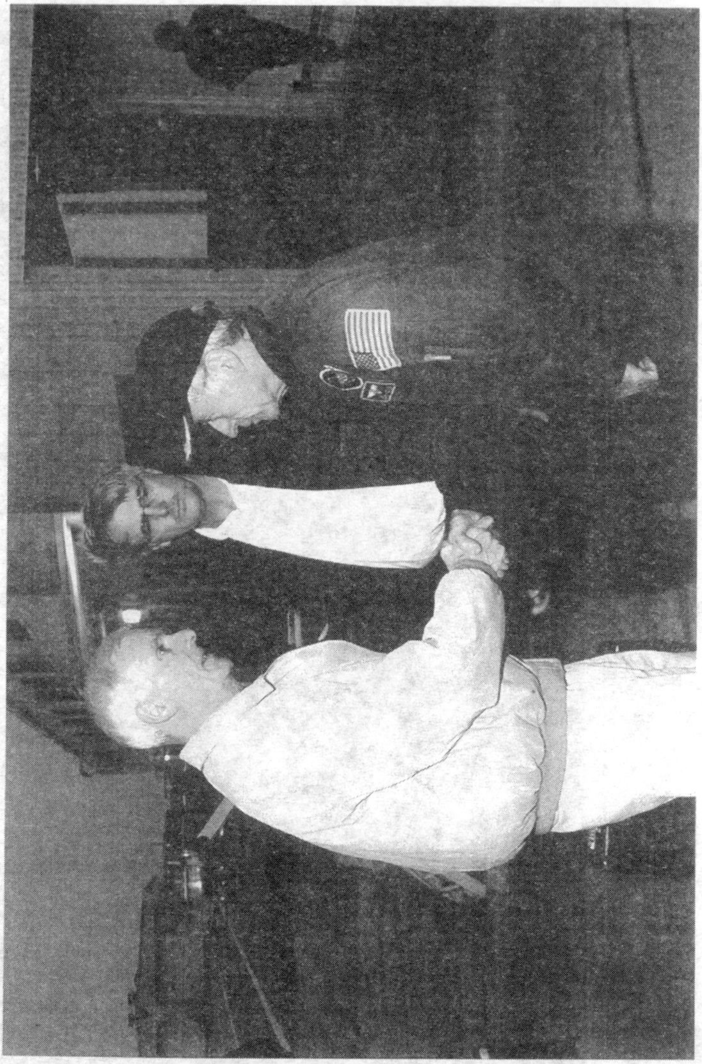

CAIB on Site

Columbia Recovery Efforts

Picture of Debris at Barksdale

Aerospace Safety Advisory Panel 2001 Annual Report

"NASA and its contractors maintain excellent safety practices and processes, as well as an appropriate level of safety consciousness. This has contributed to significant flight achievements. The defined requirements for operating at an acceptable level of risk are always met."

Aerospace Safety Advisory Panel
Annual Report for 2001
Page 9

NASA's Vision
To improve life here,
To extend life to there,
To find life beyond.
NASA's Mission
To understand and protect our home planet
To explore the Universe and search for life
To inspire the next generation of explorers
...as only NASA can.

Chairman BOEHLERT. Thank you very much, Mr. Administrator.

The Chair recognizes the Ranking Member of the House Science Committee, the gentleman from Texas, Mr. Hall.

STATEMENT OF HON. RALPH M. HALL, U.S. REPRESENTATIVE FROM TEXAS

Representative HALL. Thank you, Mr. Chairman, and thank you, Chairman McCain, and I thank this group.

Mr. O'Keefe, I thank you, your Deputy and your Associate Administrator for Space Flight and those valiant people who sit behind you there that contribute so much day in and day out. We're grateful to you.

And I speak for Bart Gordon, who is the Ranking Member of the Space Subcommittee, who has the same respect I have for the leadership. And this is a day and time when we should be neither Republicans nor Democrats, but Americans. And I think it's a day in time when we come together.

And, Mr. Administrator, you did a great job that Monday, that fateful Monday, in Houston. Thank you for that.

I think, certainly, that this one of the most painful hearings that I've ever had the duty to try to get prepared for. It's less than two weeks now since the Shuttle broke apart in the sky up over my home in my area in Texas. I'm saddened every time I think of these seven brave astronauts and the grief-stricken families that they left behind. I knew three of them very well.

And the young lady from India, who had accomplished so much and came so far, came to my district on more than one occasion, had a great sense of humor, was really great for the program. In one of her speeches to one of the classes in Canton, in Vanzant, Texas, one of the students said, "We have a hard time pronouncing your name." She said, "That's all right. I have a hard time pronouncing yours."

(Laughter.)

Representative HALL. She had a way with youngsters and was very helpful.

I know that there are a lot of questions about what went wrong, and I'm going to shorten my speech here because we have so many others that really should be heard from and we have questions that we have to ask you.

There has also been a lot of speculation as to what or who may be to blame for the accident. The reality is that it doesn't appear that anyone yet knows what caused the accident, although the NASA Administrator may have some information in the progress of the investigation to share with us here today. And you've done that, and I thank you. And I think the questions will elicit more information and will be helpful to us.

So it might be some time before we'll be clear on what factors have contributed to the accident. Thus, it's important that we have a thorough and, I want to stress, independent, as Mr. Gordon has stressed, investigation of the accident so that the American people can be assured that everything's on top of the table. And I know that's what everybody in this room wants. Anything less would be a disservice to the courageous men and women who died on the *Columbia*.

Our nation is grieving. We're mournful at this time. And the families are in mourning. But time lessens and sometimes heals that. But that same time is going to bolster the need for an independent investigation, and that's what we're looking for. And, Mr. Administrator, I understand that you've pledged to do that, and we thank you for that.

I think we need to take a very close look at what can be done to improve Shuttle crew survivability. As a long-time Member of this committee, I've always had problems cutting the NASA budget, because not having the knowledge that you men have, not having the exposure of life or death that so many of you have, I didn't know how to cut it or how to recommend cutting it without endangering someone. So we've had to call on the Administrator to do that. We had to call on Dan Goldin to do that. He did it, and, I think, did it in a good manner. We call on you, Mr. O'Keefe, to give us that same type leadership, and we pledge our support to you as we seek out causation and how to keep it from ever happening again.

NASA's talking about spending upwards of $9 to $13 billion, by its own estimates, over the next decade to develop a still-to-be-defined Orbital Space Plane. That's long-range planning. We have to have that, and we have to have some short-range decisions.

I think we need to examine whether some of that money would be better spent on developing crew escape systems for the existing Shuttle fleet and on completing a simply, reliable U.S. crew rescue vehicle for the International Space Station, and doing both of these things as soon as possible.

With that, Mr. Chairman, I'd like unanimous consent to put my full speech in the record, and I yield back the time.

Thank you.

Chairman McCain. Without objection.

[The prepared statement of Mr. Hall follows:]

Prepared Statement of Representative Ralph M. Hall

This is one of the most painful hearings that I have had to prepare for in all my years in Congress. It is now less than two weeks since the Space Shuttle *Columbia* broke apart in the sky over my home state of Texas. I'm saddened every time I think of those seven brave astronauts and the grief-stricken families and friends that they left behind. They made the ultimate sacrifice for the cause of space exploration, and we shall miss them dearly.

When the STS–107 mission was launched into orbit in mid-January, I was looking forward to what we would learn from it. As many of you know, it was a mission dedicated to research. As a result, it was a mission that offered the promise of improving the lives of our citizens back here on Earth. That is the vision I have long had for our space program: learning things in space that can be used for the benefit of all Americans. It is what the International Space Station should be about if this nation will step up and honor our long-standing commitments to complete the project. And it is what the astronauts of STS–107 were trying to accomplish on their ill-fated mission.

I know that there are many questions about what went wrong. There has also been a lot of speculation as to what or who may be to blame for the accident. The reality is that it doesn't appear that anyone yet knows what caused the accident, although the NASA Administrator may have some information on the progress of the investigation to share with us today. So it's likely to be some time before we can be clear on what factor, may have contributed to the accident. It thus is important that we have a thorough, independent investigation of the accident so that the American people can be assured that nothing is being hidden. Anything less would be a disservice to the courageous men and women who died on *Columbia*.

Whatever the specific cause of the *Columbia* accident, we in Congress need to take a hard look at where we go from here. NASA's latest proposal doesn't envision having an alternative means of launching crews into space for another decade or more. And in any event, NASA seems to lie committed to flying the Shuttle to the Space Station throughout the lifetime of the Station. A decade or more is a long time. If, God forbid, there is another accident sometime during that decade, will we be able to look back and say we had done all we could to improve the crew's chances of survival? I hope so.

For example, the Aerospace Safety Advisory Panel's March 2002 report found that *17 years* after the *Challenger* accident the Shuttle program still is facing a situation where:

> *"there is no in-flight crew escape system for the Orbiter other than for abort below 20,000 feet during a controlled glide"*

and it recommended that NASA:

> *"complete the ongoing studies of crew escape design options and implement an improved system as soon as possible."*

I think we need to take a close look at what could be done to improve Shuttle crew survivability. NASA is talking about spending upwards of $9 to $13 billion by its own estimates over the next decade to develop a still-to-be-defined Orbital Space Plane. I think we need to examine whether some of that money would be better spent on developing crew escape systems for the existing Shuttle fleet and on completing a simple, reliable U.S. crew rescue vehicle for the International Space Station—and doing both of those things as soon as possible. I don't think the brave men and women who serve in our nation's space program should be needlessly put into harm's way any longer than necessary if there are practical remedies available.

Thank you, and I yield back the balance of my time.

Chairman McCAIN. Senator Stevens has to return quickly to chair the conference concerning the Omnibus Appropriations bill, which all of us eagerly await the result of his work, and so he'd like to make a brief statement.

Senator Stevens.

STATEMENT OF HON. TED STEVENS, U.S. SENATOR FROM ALASKA

Senator STEVENS. Mr. Chairman, I thank you, and I do have to return to that conference. I have come because the gentleman that's before you I consider to be one of the closest friends I have in the world. I think members should know who he is.

He came to Washington as a White House fellow. He worked for the Department of Navy, and then he became a Senate employee and became the chief of staff of the Defense Appropriations Subcommittee. He went from there to become the Comptroller of the Department of Defense, the Secretary of the Navy. He subsequently became a professor at Syracuse University, then a deputy director of the Office of Management Budget, and is now the administrator of NASA.

I know of no one who has committed himself to good government and conducted himself in the finest of our traditions than Sean O'Keefe. He is a man of integrity, of complete honest and openness in all he does. I would back him, as I know he would me, with my life. And I urge you to listen to Sean O'Keefe today. He'll tell you the truth.

Thank you very much.

Mr. O'KEEFE. Thank you, Senator.

Chairman McCAIN. Thank you very much, Senator Stevens.

DISCUSSION

EFFECTS OF BUDGET DECISIONS ON SHUTTLE PROGRAM

Thank you, Mr. O'Keefe, for your presentation.

Look, one of the issues that is going to be talked about a lot today by a lot of the members is the issue as to whether the NASA's budget was, "starved," cut to the bone. There will be allegations that certain recommendations were made by certain people.

For example, the Aerospace Safety Advisory Panel annual report for 2001 stated, "The current and proposed budget are not sufficient to improve or even maintain the safety risk level of operating the Space Shuttle." I've seen a lot of rhetoric in the media, and you have too, that you were "starved." That was not my experience, as Chairman and Ranking Member of the Commerce Committee, but I think it's very important that you take that issue head on and immediately.

Mr. O'KEEFE. Yes, sir. No, I fully agree. There is no question, this is a concern that we continue to have, as well, and to assure that all of the facts that are laid out on that particular matter. As it pertains the views of the ASAP and the advisory committee, as well, their reports, I think, reiterate consistently their view that the future concerns about Space Shuttle operations and safety considerations were the matter they were most focused on. As a consequence, their continued effort that I see in the report before us always is that they quote specifically, "It's important to stress that the panel believes that safety has not yet been compromised. NASA and its contractors maintain excellent safety practices and processes, as well as an appropriate level of safety consciousness. This has been—contributed to significant flight achievements in the defined requirements for operating, and an acceptable level of risk are always met."

So their concerns were always presented in the context of future approaches. And, as a matter of fact, if we call up slide number 35, that will cover that particular question, as well. Their focus was always on the future operations as well as future efforts that were to be engaged in.

At the present time, in terms of current operations and activities they certified as recently as a year ago, they felt that the current operations were concentrated on very specifically to assure flight safety as a primary paramount objective.

Chairman MCCAIN. I think you're going to be confronted with some numbers in further questioning, and I hope you will have responses to that, as well.

At a Commerce Committee September 6th, 2001, hearing on Shuttle safety, William Readdy, then Deputy Associate Administrator of the Office of Space Flight, acknowledged the challenges NASA was facing to maintain Shuttle safety in light of budgetary constraints, but, nevertheless, stated, "The safety of the Space Shuttle has also been dramatically improved by reducing risk by more than a factor of five." Later in his testimony, he said, quote, "The Space Shuttle is the safest, most capable and reliable transportation system in the world."

Mr. Blomberg, the former chief of NASA's Aerospace Safety Advisory Panel, testifying before the House Science Committee in April

2002 on behalf of the advisory panel, stated that, quote, "In all the years of my involvement, I have never been as concerned for Space Shuttle safety as I am right now. The concern is not for the present flight or the next or perhaps the one after that. In fact, one of the roots of my concern is that nobody will know for sure when the safety margins have been eroded too far. All of my instincts, however, suggest that the current approach is planting the seeds for future danger."

How do we reconcile those two statements by two highly regarded individuals who are within the bureaucracy of NASA?

Mr. O'KEEFE. Yes, sir. Well, just to clarify the record on that, Bill Readdy, of course, is the Associate Administrator for Space Flight. Richard Blomberg was a independent external member of and a chairman of the advisory panel for safety, so he was not a full-time NASA employee in that regard. He was representing a panel view.

Reconciling that is—I think if you trace the history just a little bit, the plan that had existed until this past November contemplated the retirement of the Space Shuttle fleet as early as the middle of this decade, certainly no later than 2012, so it altered over the course of several years from about '95 forward, the best I can tell, over the history of this.

Based on the recommendations of that advisory panel on safety, as well as the testimony and comments made at several different committee hearings, as well, we went back and really looked seriously at the question of what it would take in order to maintain Shuttle operations for a sustained period of time, what kind of continued upgrades would be necessary, modifications necessary, in order to assure safe flight operations, and on the basis of that, as recently as last summer, went through that planning effort, which ultimately yielded the amendment that was sent forward by the President on November 13th of last year to propose a specific change in the funding profile for Shuttle, which envisioned a maintenance of that asset for a sustained period of time, though next decade.

So the focus of these concerns, which were all exactly as you stated, Mr. Chairman, in context of future concern, were the things we were very mindful of, took heed of, made adjustments to, and specifically put in plan in order to assure that we covered those kinds of concerns in the future and addressed those.

As it pertained to current flight operations, again there was no indication that I knew of that raised concerns along the way of current flight operations. If anything, the diligence that I see among the entire folk in NASA, in the community, is very much that of a culture that's dedicated to assuring safe flight operations, or else the launch doesn't occur. And that is the mindset and ethos we continue to encourage and will continue to encourage in the future.

CHANGES NEEDED TO ASSURE AN INDEPENDENT INVESTIGATION

Chairman BOEHLERT. Mr. O'Keefe, as I mentioned in my opening statement, I remain concerned about language throughout the charter of the *Columbia* Accident Investigation Board. The language would appear to indicate that everything the board does is subject to NASA approval, and that, to me, raises some fundamental questions about the independence of the board. And we all want the board to be independent, and not just in name, but in fact.

Are you willing to re-examine that charter and remove some of the language that raises these questions and make adjustments in other places?

Mr. O'KEEFE. Yes, sir. We'll continue to work with Admiral Gehman to assure whatever he needs in order to guarantee the independence and objectivity of that board. We will absolutely work with him, without objection.

Chairman BOEHLERT. I mean, just to give you an example, the independent board will conduct activities in accordance with the provisions of applicable NASA policies and procedures. And then it goes on to say, "The interim scheduled board activities, interim board reports, and the submission of final board report, in coordination with the NASA Administrator." I would think that they would have independence, they could schedule their meetings and determine the type of report they want to submit. Of course they will submit the report to you. But the report should come also to the—the final report—not just to the NASA administrator, but to the President, the Congress, and the American people. So I think that charter has to be revisited, and very promptly.

We've already had conversations with each other, and I have had with Admiral Gehman, and both have assured me that additional members from outside the community, so to speak, and experts in different fields, will be added to the board. I think that's very important.

It's essential that we maintain the independent nature of the board.

Mr. O'KEEFE. I fully concur, Mr. Chairman. And, as a matter of fact, I think in Admiral Gehman's press conference yesterday, he was very explicit that all of the factors he needed in order to maintain independence and to be an objective investigator in this particular case, for all of his board members, was what the present condition required.

Having said that, if, on examination, the latest change that was made to the charter at his request, we made it, if he wants further changes they'll be made, as well, anything that it takes in order to guarantee their independence, because we will be guided by their findings. And, as I've reiterated publicly as well as to you, sir, and to him, that about the only thing that will be unique about the reporting requirement is that he'll be putting an address that says NASA on it, on the envelope, sending it to us. But that report will be made public concurrent with its receipt. So he will be reporting to the President, to the Congress, to the American people, to all of us simultaneously.

CONTRACTOR INCENTIVES AND OBLIGATIONS

Chairman BOEHLERT. Well, I was comforted by my conversation with Admiral Gehman, because he is insisting on independence, and he has also indicated, obviously, he'll have to have a liaison with NASA and use some of your resources, but the staffing will be independent of NASA, and I think that's essential.

Obviously, we're all going to be spending a lot of time looking at Shuttle contracts even though there's no indication at this point that they are a problem. How comfortable are you that the incentives in the current contracts captured the proper balance between efficiency and safety? And then, as a follow-on, do the contracts have clauses that will ensure that the contractors have to, must, cooperate with fully with the Gehman investigation?

Mr. O'KEEFE. Yes, sir. On the first part, the emphasis on safety in the current Space Flight Operations Contract that we have very much emphasizes the safety parameters. And, indeed, they have tremendous incentives to do better each and every time. And as a consequence of that, there are a series of weighted factors in the guidelines that heavily look at the metrics of any difficulties or problems on orbit or at launch or any other time. As they drive those factors down, they're given a specific incentive to do better in each of those cases. So they've got a powerful motivation to want to move in that direction.

In doing so, I think the approach also yields some efficiencies. But that's a secondary matter, at most. And so, as a result, there are real advantages and real emphasis on the safety considerations that are currently built into that contract framework.

As it pertains to their cooperation with the Gehman Board, positively we will advise them, and have, that we fully expect everyone to be cooperative with that board. We have absolutely nothing to hide. There is no evidence or no fact that we can think of out there, short of national security information or some private proprietary issue that some individual may want to assert, that would preclude us from making any information available. And so everyone within the contractor community should feel the same as we do.

Chairman BOEHLERT. But is there something more than a moral obligation or a desire? Is there something that binds them, commits them, to cooperate fully with the Gehman investigation board?

Mr. O'KEEFE. I will take you up on your opening statement that, on that contract clause, I don't know, but I'll find out.

[The information follows:]

The Space Flight Operations Contract (SFOC) and the External Tank (ET) contractors are required under their respective contracts to cooperate with the Gehman Board. Review of the current SFOC and the ET production and flight support contract indicates that the contractors are contractually obligated to cooperate fully with the Gehman Board or any other Government initiated space flight mishap investigation board.

Both contracts provide extensive contractor responsibility to perform or coordinate anomaly investigations or to cooperate in the event of a NASA led investigation.

The SFOC Statement of Work (SOW) elements requiring the contractor to support Space Shuttle Program mishap and accident investigations include those related to catastrophic events such as the loss of Columbia. Specifically, SOW paragraphs 1.3.1.2 – "Operations Safety" and 1.3.3.1. – "Product and Service Quality" include this requirement. The latter specifically states: "The contractor shall provide for the identification, reporting, and elimination/disposition of non-conformances/problems/anomalies/failure analysis in accordance with NSTS 07700 and NSTS 08126." In addition, SOW paragraph 1.4 – "Flight Elements" includes the requirement to support unplanned program contingencies following the requirements of NASA document NSTS 07700 (which provides for the processes followed for the Columbia accident investigation). SOW 1.1.1.4 – "Risk Management," requires government approval for disposition of all "out-of-family" anomalies involving, such as here, first time occurrence of a failure mode, affect on performance or reliability of safety or mission success critical hardware functions, unexplained anomalies, or government design element analysis or assistance.

The foregoing specific references provide a clear obligation to perform extensive anomaly analyses as necessary to determine the cause of malfunctions and resolve the problem to maintain program safety. The additional obligation to cooperate with NASA-led anomaly investigations is also clear. NASA appointed the Gehman Board in accordance with contract reference document NSTS 07700 as its method of conducting the flight anomaly investigation. It follows that the contractor is obligated to cooperate with the Gehman Board, as it would be required to cooperate with any other NASA mishap investigation.

SFOC Article H.19, "Access to Contractor Data" further supports the obligation. It states, "The Contractor shall provide the Government unimpeded access to all areas determined by Government representatives as necessary for surveillance, audit and independent evaluation purposes." SFOC Article H.25, "Government Insight," provides additional cooperation requirements, and states, "The Government may schedule fact-finding meetings with the Contractor and subcontractors as necessary to discuss issues requiring Government insight. Scheduling and format of these meetings shall indicate whether exchange of information will be required, and the number and expertise of Contractor/subcontractor personnel who shall attend the meetings. When requested by the Contracting Officer or designee, the Contractor and subcontractors shall provide necessary support to the Government when it audits the Contractor or subcontractor and for the Government-Contractor/subcontractor meetings." These types of activities would be involved in cooperation with the Gehman Board. The clause further provides for "technical visibility into the Program, maintained through audit, surveillance, assessment of trends and metrics, software independent verification and validation, the flight readiness review process, and review or independent assessment of out-of-family anomalies occurring in any phase of the program."

The External Tank contract also includes specific SOW provisions for anomaly analysis. Contract Attachment J-12, "External Tank Flight Operations," paragraph 2.2, "Flight Evaluation," states, "Appropriate anomaly investigation will be performed by the Contractor." Although the language is limited to "the period from countdown (T-1 day) through ET breakup," this is broad enough to cover any flight events related to the External Tank flight performance before its breakup, including reentry events that occur long after ET breakup. SOW 4.9.4, "External Tank Launch Support Services (KSC)" (WBS 3.6.2.4), states the contractor "will support the Government and the Shuttle Processing Contractor (SPC) with launch site expertise on ET design and processing requirements. The Contractor will also provide or coordinate support in the areas of . . . anomaly resolution." This requires support to and cooperation with the Gehman Board as a NASA chartered anomaly-investigating organization.

A further contract obligation to cooperate is provided by contract Attachment J-1 provision 4.9.2, "Flight Support (WBS 3.6.2.2)," that states, "The Contractor shall perform tasks in support of the External Tank Flight Support as identified in Attachment J-12 of this contract beginning January 1, 1989. Any extraordinary support resulting from major anomalies or changes in mission model will be the subject of separate contractual action." Although an equitable adjustment to compensate the contractor for "extraordinary" support may arise, this very specifically indicates an obligation to cooperate with any NASA anomaly investigations as directed.

Finally, the contemporaneous conduct of the contractors in performing in a cooperative role, immediately and without any question, may provide a compelling indicator of contract obligation. By all reports, the firms involved are already fully cooperating with the Gehman Board. Whether performed as a result of specific provisions or the general obligation of a contracting party to cooperate in accomplishing clearly defined contract objectives, the conduct of the parties may be an overriding factor in determining contractual obligations.

The foregoing relies on review of the current SFOC, NAS9-20000, held by United Space Alliance, LLC (a joint venture co-owned by, The Boeing Company and the Lockheed-Martin Company), and the ET production contract, NAS8-36200, held by the Lockheed-Martin Company, under which the ET used for Columbia's final flight was delivered.

Chairman BOEHLERT. Thank you very much.
Mr. O'KEEFE. Thank you, Mr. Chairman.
Chairman MCCAIN. Senator Hollings.

NASA'S BUDGET REQUEST TO OMB

Senator HOLLINGS. Right to the point with respect to costs, and I'm sure you don't have, Mr. O'Keefe, the actual figures with you, but on the shortchanging of the space program, we had a report yesterday in the *New York Times* stating that we cut the space program $800 million. I've been checking it overnight. At my check, it's $700 million.

Be that as it may, what's the truth? That's what we want. Last week I asked Mitch Daniels, Director of the Office of Management and Budget, to furnish the Budget Committee the figures of what was requested by NASA. I want you to furnish the figures that were requested by NASA of OMB for the last 10 years. And not just this Administration, but the past Administrations so we can see the trend line and exactly how we financed it.

Yes, we all are trying to find out the cause. But, in the meantime, as you try to piece together the retrieved parts and everything else like that, I don't know how long that'll take, a year or months, whatever it is, we all want to see the space program continue. And for this senator, I don't want it to continue with upgrades.

I've heard enough about these upgrades. We've lost 14 astronauts and $5 billion in hardware with upgrades, and we had a new reusable launch vehicle. We had a spaceship that was cancelled the year before last. Then we had a Reusable Launch Vehicle, and that was cancelled last November, just a couple of months ago. And you said we were going to use these vehicles, Shuttles, until 2020. Are you willing to use one of these Shuttles with all of the tiles flying off? And after all of these losses, you'd still want to use them until 2020 and not get new technology, Mr. Administrator?

Mr. O'KEEFE. Yes, sir. No, thank you. As I understand the history here, the effort during the course of the '90's was in anticipation of a retirement of Shuttle concurrent with the introduction of a new Reusable Launch Vehicle. That was envisioned to be the X–33, as I understand it. Based on a variety of technical issues, which were based on the assumption that a series of unconquerable engineering and laws of physics challenges would be overcome, ultimately that—two-plus years ago, the choice was made to cancel that program and to continue with Shuttle operations.

What we proposed a year ago and is not a cancellation of any RLV effort, Reusable Launch Vehicle, instead it's a selection, if you will, of looking at the Integrated Space Transportation Plan.

What's included in the November amendment that the President sent forward for the 2003 budget is a selection specifically of an Orbital Space Plane option which, frankly, is not a technology leap. It is the capability of putting aboard an Expendable Launch Vehicle, a orbital space system, space plane, that would be launched in a conventional manner using an Expendable Launch Vehicle.

The next generation beyond that is what we focus on our Next-Generation Launch Technology. So we've tried to narrow and focus a lot more the Space Launch Initiative efforts in order to get some near-term gain to supplement, to complement, the Space Shuttle and to provide that dynamic as well as flexible return system and transfer system to the International Space System and also to pursue the development of a Reusable Launch Vehicle that may be, hopefully, the product of breakthroughs that were not possible that forced the motivation or the cancellation a few years ago of the X–33.

So we're trying to do both of those concurrently, but to get some near-term capability, and, in the meantime, use Shuttle in the future as a cargo lift, heavy lift capacity, which is what it was really designed to do in the first place, rather than a crew transfer capability. So we're trying to balance both ends of that to utilize capabilities for their best purposes as we move along.

Senator HOLLINGS. We've got to find out what you think we ought to Appropriate. We all want to continue space exploration, but we just don't want to waste time waiting on the results, on the one hand, and trying to find out what we already know. Let's get on and get your best advice on how we should proceed to get going on this thing, and not just with upgrades.

Mr. O'KEEFE. Yes, sir. No, the——

Chairman BOEHLERT. Thank you very much, Senator.

Mr. Hall.

CREW ESCAPE SYSTEMS

Representative HALL. Mr. O'Keefe, you heard my opening statement. And I'm, quite frankly, disappointed that 17 years after the *Challenger* accident so little attention has been given to developing crew escape systems for our astronauts, whether they're flying on the Shuttle or whether they're in the space station. I know you share that.

I share with you the disappointment and the blame—I've been here 23 years, so it's a partnership for us, and that's what it is, that's what it'll continue to be, but especially since NASA has indicated that the Shuttle is going to fly for another decade and a half, and maybe, probably, longer than that, and in light of a media report, I think on February the 5th, that NASA's most recent effort in that regard was a $5 million so-called "study" in 2001.

To put that study in context, the amount expended on that study of potential a Shuttle crew escape system amounted to just a little bit more than one-tenth of one percent of a single year's budget. That doesn't strike me as being very aggressive in your effort to look for ways to improve the odds of survival for astronauts in the event of a Shuttle accident.

That leads me to say that I have no doubt that it's going to be challenging to develop practical crew escape systems for the Shuttle, but NASA is in the business of performing miracles, NASA is in the business of meeting challenges, and we call on you to do that.

I'm very afraid that a clue as to why NASA has not done more is found elsewhere in that article, namely, and I quote, "The proposed fixes were also seen as prohibitively expensive additions to an already aging and financially strapped Shuttle fleet." We've seen a similar situation with regard to developing a Crew Rescue Vehicle for the International Space Station.

So, as you know, NASA decided to cancel the ongoing development of U.S. Crew Rescue Vehicles, just a demonstration vehicle, we thought was nearing its flight test. And now we're dependent on the Russians for their Crew Rescue Vehicles until the end of this decade.

So, in the meantime, I guess my question to you is, Did you explicitly consider investing in the development of Space Shuttle crew escape systems when you revised the Integrated Space Transportation Plan that you announced last November?

Mr. O'KEEFE. Yes, sir. We're continuing to look at what we would use as enhancements, if you will, of the Space Shuttle as part of that November amendment that was submitted last November, and we're getting together here, had planned to all along, to identify that priority set of what will emphasize the highest safety margin improvements that could be yielded from different modifications of the orbiter system.

But with regard to the specific crew escape efforts, recall that since *Challenger* there have been a number of operational changes made. There is an egress system that was put into place right after the *Challenger* accident that was part of the Rogers Commission recommendations—that ultimately stemmed from it, I should say—that we put in to place that now still exist to this day.

Once launched, though, there is a number of different approaches that have been proposed, examined, reviewed, and all of which added significant amounts of weight, I'm advised, to the overall effort, and so, as a consequence, were viewed to be technically infeasible.

What we have instituted, though, is, again, a series of abort procedures. And, as recently as the December launch of the STS–113, on a perfectly clear night at Cape Canaveral in Florida, where everything was nominal, everything was ready to go, all the systems were completely operational, we scrubbed the launch because the alternate abort site at Zaragossa, Spain, the weather continued to be marginal. So we take every precaution in this process in order to assure that, all the way through assent, that every possible opportunity is there as much as possible.

But, again, the idea of an escape system was looked at, examined very thoroughly, and the conclusion was that the weight factor would almost be prohibitive in terms of its technical clarification.

So we'll continue to look at that. We'll go back and look at it again, you bet. In light of this circumstance, we really do need to focus entirely on what all the alternatives are, and I guarantee you, sir, we'll make that part of our effort underway now as part of this November amendment that is before the Congress to consider for the 2003 program, that we'll factor that into the equation and proceed as appropriate.

Representative HALL. Even on a local level, city councils rarely ever fix a bad bridge or a bad turn in the road until a teenager gets killed, and then it's too late. It seems to me that we launched the vehicle without any ability to dock. We lost the vehicle because we didn't have telescopic ability to inspect. We have three birds left. I just urge caution. And I thank you for your time.

I yield back my time.

Mr. O'KEEFE. Thank you, Congressman. I appreciate it very much.

Chairman MCCAIN. Senator Brownback.

QUESTIONING NASA'S GOALS AND OBJECTIVES

Senator BROWNBACK. Thank you very much, Mr. Chairman. And, Mr. O'Keefe, thank you very much for coming in front of our Joint Committees here and your great leadership at NASA in a very difficult and trying and challenging time.

We all grieve the loss of human life that's happened to NASA. And at this particular point, I'm chairing a subcommittee in the Senate that'll be dealing with this, and I want to work with you and your agency as we lay out the future of manned space flight in the United States.

I want to ask you about the broad objectives and broad program objectives that we're talking about right now. It seems to me that the space program is really at a critical juncture and that the totality of the space program is. And it's got to decide amongst a couple of competing options. One is to maintain the current set of programs and current missions. The second one that you read a lot about in the newspaper and people speculate is to dramatically reduce manned space flight, go into more robotics and different types of vehicles, questioning about the safety and to try to be more safe. And the third, a number of people are saying that our vision is too small in space currently right now, that we need new initiatives, we need to go back to the Moon, we need to go to Mars. And we're at this tremendous fork-in-the-road decision of which path to take.

It's certainly my intent in the Subcommittee to look closely at where NASA has been and where you are today and where we plan to go into the future. And, most importantly, we need to discuss, as well, the financial situation, the terms of how we get NASA where it needs to be. I'm glad to see that, in the Appropriations Committee, we're putting in an additional $414 million over the President's request.

The goal is to reflect an accurate and effective determination for the future of NASA, and I would simply like to ask you, What have you done recently—and I realize you're dealing with the tragedy mostly now, but—to look at this need for a review of the mission of NASA amongst these three policy objectives, have you had a chance to start contemplating some of that? And I hope you'll be open to working openly with the Congress as we look at this fork in the road we're in right now.

Mr. O'KEEFE. Yes, sir. Of course, Senator, I'd be delighted to work with you and any other Members of Congress to sort through really what is the proper role and objective of NASA in our pursuit of exploration objectives, always.

We have, indeed, had an opportunity prior to February 1st to really think seriously about what is the strategy and the focus of how we concentrate on what we do best, and do that only in a way that guarantees and assures that we—to offer to folks that we can actually deliver on and have a capability to look at longer-term exploration objectives. And I think that's contained in the strategic planning documents that were all forwarded, along with the budget that was submitted by the President just last Monday.

Prior to that, in all the efforts we put into it, was to think seriously about the very kind of questions you've raised, and, again, to narrowly focus on the kinds of exploration and discovery objectives we think we do exceptionally well. And, for those that are done by others or can be pursued elsewhere, to leave that to folks who have expertise or capabilities that would otherwise have to be duplicated by us.

What it leads to, I think, is a stepping-stone approach, an exploration strategy, if you will, that assumes that we start off with a series of robotic capabilities, and moves forward then, thereafter, toward other exploration objectives that may or may not involve human involvement.

The best example that we've seen played before us in the last several years is the Hubble Space Telescope. There was a $2 billion capability that, when launched in 1992, in fairly short order was deemed to be, as a marvelous robotic capacity, a capability that was in need of an eye examination, if you will, a lense correction. And it was, at that time, determined to be a $2 billion piece of space junk. A year later, we were able to send a Shuttle flight with astronomers and other engineering capability that was resonant among the astronauts there to make that correction. That could not have been done remotely.

And so the human intervention that was necessary to adjust that, and all the servicing missions we've done since that time, have yielded the kind of astronomy breakthroughs and discovery, just in this past year, that we never dreamed imaginable. So that combination, that heel-toe kind of approach towards a strategy that utilizes robotic capabilities, much like we're going to do here in the coming months when we send the Mars explorers in May and June of this year intended for landing in January of '04, is to then consider all the efforts we've got to do to prepare for, then, the follow-on kinds of exploration objectives.

But, again, the reiteration of the first commitment to you, positively we'll continue to work together and refine this strategy to assure that we do it with least risk, but the greatest opportunity for exploration and discovery potential.

Senator BROWNBACK. Thank you. And I look forward to working with you on that design of where we——

Chairman MCCAIN. The gentleman's time has expired.

Mr. O'KEEFE. Thank you, Senator.

Chairman BOEHLERT. Mr. Rohrabacher.

Representative ROHRABACHER. Thank you very much.

First and foremost, I'd like to associate myself with the remarks and the concerns of Chairman Boehlert about the basic nature and the importance of the integrity of this commission's investigation and our oversight of that commission's work.

Second of all, I would like to just note that, at the memorial service down in Texas I was touched particularly by the people of Texas, and especially by the children of Texas, who, as we went to this memorial service, they came out on the streets and roads and waved little American flags and had little signs up to encourage us, and it was very encouraging for our country. So we recognize that there was a special bond between the children of America and our astronauts.

And today we're building, and we're going to make sure that we build a better future for our children. And if there's going to be a better future for our children, we've got to have a viable space program that will keep them in the forefront of this great human endeavor of going into space and pioneering space.

So let me—I have a few thoughts, and I'll have a few questions for you. The hardworking and patriotic people of NASA have always understood and appreciated the risks with space exploration, especially manned missions. Unfortunately, in the past 17 years, we have been reminded of the dangers of human space flight with the destruction of the Space Shuttle *Challenger* in 1996—or '86, I should say —and now the *Columbia*.

Seventeen years ago, we took a step backwards for a few moments to take a look at that tragedy and pinpoint to our satisfaction what caused it and then correct those causes, at least the technical causes of the loss.

Today, I am confident that Admiral Gehman and his commission will get to the truths that will help us understand *Columbia*'s fatal accident. However, many questions need to be addressed that transcend the immediate technical and managerial problems of this tragedy.

We're going to hear a lot about the technical end of it, but there's a lot of other questions that go way beyond that. The lack of long-term goals or a unifying vision for America's space effort, for example, needs to be addressed to fully understand this tragedy. This failing, I believe, weaken the efforts that would have been taken to replace the Shuttle system long before age became a factor. And we will find out, I believe, that age was, indeed, a factor.

Perhaps Mr. Hollings, or Senator Hollings, is right, perhaps it's simple what we're looking at. Perhaps it was the tiles and—in terms of a simple answer; and a more complex answer might be facing—it might be age. But this tragedy and this investigation, nonetheless, gives us an opportunity to revisit the fundamentals and make recommendations that will chart America's future space endeavors.

NASA's leadership has faced, and will continue to face, the challenge of exercising responsible stewardship with limited resources while providing a coherent blueprint of what can be accomplished and how it can be accomplished. But hopefully, forward-looking strategies will lead us to incremental advances that will then permit us to achieve long-term objectives. A new propulsion system might be a good start, as well as, perhaps, a look at robotics and remote control on the part of NASA, a new commitment on that end. But before we move forward, we must fully understand why these seven people perished.

My question to you today is, the age factor, Was this Shuttle's age, a 30-year-old system, a major factor in this tragedy we're investigating today?

Chairman BOEHLERT. Administrator O'Keefe.

Mr. O'KEEFE. Well, thank you, Congressman. And, again, I associate myself entirely with your observation that ultimately the investigation will be what guides us to that conclusion. And if that is a factor, you bet, that's exactly what we'll operate on.

Having said that, it is worth noting two really important factors on this. The *Columbia*, indeed, is the oldest, or was the oldest, of the four orbiters. It was delivered in late 1970's. Its first flight was in 1981. It was half the age of the average bomber aircraft that flew in Afghanistan just a year ago that prosecuted that very important effort that we were engaged in there.

So the air frame condition on this—each time we go through this orbiter major modification I referred to in the opening statement—is essentially the equivalent of the same kind of effort the military, the Defense Department, goes through of tear-down of every single element of the capability of the orbiter itself, its structural integrity inspected and examined very carefully, then rebuilt to modernize it to contemporary capability. That particular effort had just been conducted, an 18-month tear-down of the Shuttle *Columbia*, and delivered early last year in advance of the March flight that went to Hubble, that did the servicing mission. STS–109 was the first flight of *Columbia* after that particular Orbiter Major Modification effort, which, again, is patterned very much after the depot kind of approach that's taken at all of the Defense Department-related assets, only even more exhaustively to conduct the upgrades. This was the second flight after that particular tear-down.

So the age factor, again, you're exactly right, the investigation may yet prove or may demonstrate to us that there was a contributor there. But in terms of our efforts to assure that not be a factor, again, it appears to be every element of diligence could be done to assure that, there was a previous flight that operated just perfectly, no difficulty whatsoever on *Columbia*, no structural defects upon return. And upon every single orbiter flight return, we examine all of the elements of the Shuttles themselves, the orbiters themselves, and we move it through the Orbiter Processing Facility to assure that any damage, any structural problems, anything are detected. And there was absolutely nothing wrong with the *Columbia* that we could detect in that regard. So when it flew on its second flight, it was in the same shape it was when it left the Orbiter Major Modification program just a year before.

HISTORY OF TILE DAMAGE AND LOSS

Chairman MCCAIN. Senator Breaux.

Senator BREAUX. Thank you, Mr. Chairman.

Mr. O'Keefe, thank you. I want to also congratulate you on the method in which you've handled this tragedy and the openness that I think we've seen from NASA in how you are approaching the investigation, both internally and with the external investigation, as well.

Let me ask, do we have any idea of how many times the insulating tiles have come off a Shuttle during launch and how many tiles have actually come off during the history of the Shuttle launches?

Mr. O'KEEFE. To the best of my recollection, sir, in our efforts there, it's no more than a half a dozen times that was specifically related to it. After each flight, there is always, again, as I mentioned just a moment ago to Congressman Rohrabacher, an assessment summary that's conducted to look at each element of the orbiter when it returns. There's also an inspection of the external tank, which, as you're aware, when it reaches the upper atmosphere, it disintegrates. The two Solid Rocket Boosters, once expended, drop back into the ocean——

Senator BREAUX. Well, but on the tiles themselves.

Mr. O'KEEFE. I'm sorry.

Senator BREAUX. How many times have the tiles come off, and how many tiles have come off during the history of the Shuttle launches?

Mr. O'KEEFE. Yes, sir. At each flight, there's typically a ding or a scratch or whatever else from all the various efforts that occur as they are re-entering, as well as on ascent. I'll provide, for the record, a full summary of all of the times on each flight that a tile has been missing or lost or whatever else. But it was never considered to be significant, in terms of its safety-of-flight consideration that we've examined on the orbiters when we moved it through the Orbiter Processing Facility to look at the condition of the orbiters after each flight. But we'll submit that for the record, sir.

[The information follows:]

TILE LOSS SUMMARY
STS-1 through STS-113

Mission: STS-1
Vehicle: OV-102/Columbia
Number of Tiles Lost: 16
Location: Right Orbital Maneuvering System Pod
Cause: Tiles were undensified diced tiles

Mission: STS-3
Vehicle: OV-102/Columbia
Number of Tiles Lost: 40
Location: Upper Forward Fuselage and Body Flap
Cause: Scotchguard used for waterproofing adversely affected Strain Isolator Pads (SIP)

Mission: STS-4
Vehicle: OV-102/Columbia
Number of Tiles Lost: 2
Location: a) One, Aft Bulkhead; b) One, Upper Forward Fuselage
Cause: a) Bad bond; b) Hardware failure of carrier panel stud attachment

Mission: STS-41G
Vehicle: OV-099/Challenger
Number of Tiles Lost: 1
Location: Left hand Chine
Cause: Bond Failure

Mission: STS-61C
Vehicle: OV-102/Columbia
Number of Tiles Lost: 1 tile and 1 diced tile segment
Location: a) Star Tracker; b) Payload bay
Cause: a) Bond Failure; b) Diced tiles design issue (Diced tiles are no longer used)

Mission: STS-27
Vehicle: OV-104/Atlantis
Number of Tiles Lost: 1
Location: Right hand Chine
Cause: Debris impact on ascent

TILE LOSS SUMMARY
STS-1 through STS-113

Mission:	STS-38
Vehicle:	OV-104/Atlantis
Number of Tiles Lost:	1
Location:	Upper body flap
Cause:	Interference with structure
Mission:	STS-52
Vehicle:	OV-102/Columbia
Number of Tiles Lost:	1
Location:	Base Heat Shield
Cause:	Failure immediately above densification layer
Mission:	STS-68
Vehicle:	OV-105/Endeavour
Number of Tiles Lost:	2
Location:	a) Window carrier plate; b) Upper Body Flap
Cause:	a) Hard spots on SIP resulted in loss of SIP flexibility b) Cause unknown
Mission:	STS-72
Vehicle:	OV-105/Endeavour
Number of Tiles Lost:	2
Location:	Body Flap
Cause:	Cause unknown
Mission:	STS-103
Vehicle:	OV-103/Discovery
Number of Tiles Lost:	1
Location:	Inboard Elevon
Cause:	Bad bond

In the history of the Program, 69 insulating tiles have been lost during flight.

Senator BREAUX. If engineers on this particular case had determined that insulating tiles had, in fact, departed the Shuttle at some point and that it was in an area that was important and very key, could the angle of attack on the re-entry of the Shuttle have been adjusted to deflect the heat?

Mr. O'KEEFE. That's a potential maneuvering capability. But, again, there are more than 4,000 sensors aboard each Shuttle orbiter, and if there were any indication that there were any abnormalities as a consequence of tile loss or whatever else, they likely would have shown up during that 16-day orbiting mission.

More importantly, during the course of that time, recall that in each orbit you're looking at a sunset and a sunrise every 90 minutes, which means every 16 times a day, the temperature variation on an orbiter or a Shuttle flight varies by as much as three to four hundred degrees, plus—200 degrees during the sunlight, and minus about 150-plus during the darkness period of that 90-degree rotation—or 90-minute rotation each time it orbits. So that wide range of temperature variation, if there had been exposure, almost certainly would have shown up on one of those 4,000 sensors that are aboard the Shuttle orbiter to have given us an indication.

The fact of the matter remains, there were no abnormalities that would suggest that problem until 8:53 the morning of Saturday, February the 1st.

Senator BREAUX. But is there no way that these sensors or any other methodology would have determined if any of the tiles had departed the Shuttle during the actual mission, before it returned to Earth?

Mr. O'KEEFE. We don't think so. Every effort that were made on previous flights to examine any structural damage or change or whatever else using any kind of visual capabilities were either inconclusive or not of the level of granularity that really gave us that much detail. And, again, all the supporting data would have also suggested if there were problems on it.

Again, the reality remains, over that 16-day mission—and, again, the investigation may find some data that we're not aware of right now, because everything was locked down within a half an hour after the incident. If there's something else that emerges to suggest to the contrary, we're going to get to the bottom of it. But all the information we have now and after the flight and after the examination of it, suggests no abnormalities that would have pointed in that direction at all.

Senator BREAUX. What degree of certainty——

Chairman MCCAIN. The gentleman's time has expired.

Chairman BOEHLERT. Mr. Gordon.

REITERATING THE NEED FOR AN INDEPENDENT INVESTIGATION BOARD

Representative GORDON. Thank you, Mr. Chairman. And welcome, Mr. O'Keefe. I'm glad you joined us today.

Before I move to discuss other issues, I just want to stress my concern that the investigation of the *Columbia* Space Shuttle accident ultimately must be perceived as objective and independent if Congress, the President, and the American taxpayers are going to reach a consensus on how to move forward with our space program.

It's no reflection on you or the Admiral, but that's not going to be possible if there are lingering questions regarding the independence of the board.

As you know, I've raised this question for several days now. And having checked with your office this week and the NASA Web site this morning, there seems to be a clear disconnect from your statements about the board's independence and the rules you're laying down for the board.

Let me quote just a few examples of your rules, as Mr. Boehlert had earlier. The current board not only includes NASA employees, but you also require it to be staffed by NASA employees who will help write the board's final report, which goes to you. You require that the board must, and I quote, "schedule board activities, interim board reports, and submit the final board report in coordination with the NASA administrator in accordance with the applicable NASA policies."

Now, Mr. O'Keefe, I am afraid this will not pass anybody's smell test of independence. So please, let us move forward in a concrete way and put this bipartisan concern behind us.

ASAP'S SAFETY CONCERNS

Now, let me turn to another issue that is troubling. As you know, there have been numerous warning flags regarding the health of the Shuttle program in recent years. Just a few examples. April 2002, Richard Blomberg, head of the independent Aerospace Safety Advisory Panel, issued a blunt warning when he testified before this House Subcommittee. And I quote, "In all the years of my involvement, I have never been as concerned for the Space Shuttle safety as I am now."

A month earlier, the Aerospace Safety Advisory Panel gave you a report that stated, and I quote, "The current and proposed budgets are not sufficient to improve or even maintain the safety risk level of operations for the Space Shuttle."

Yet in spite of these warnings, you sent over a NASA budget request that cut the budget for Shuttle upgrades by $500 million, even while finding a billion dollars for new initiatives.

Because of my concern, I asked Fred Gregory, who was then the NASA Associate Administration for Space Flight, the following question at this same April 2002 hearing. "Mr. Gregory, how would you support the space station in the event you lost a Shuttle and the rest of the fleet was grounded for some period of time?" Mr. Gregory responded, "There would be no way to do that."

Now, I assumed that such an admission would have kicked off an intense effort to develop a contingency plan for supporting the space station. However, at your press briefing Monday, you indicated that over the next few weeks NASA would be working with the space station international partners to come up with a plan. You reiterated that earlier today.

Given the fact that you've had numerous warnings and you knew the Shuttle was grounded for two-and-a-half years after the loss of *Challenger,* I would assume Mr. Gregory's admission nearly a year ago would have been a wake-up call.

So my questions are, Did NASA prepare a contingency plan for the space station last year? If so, what was in the plan, and why do you now believe that you need to redo it? And, finally, if you didn't have a plan, why not?

Mr. O'KEEFE. Yes, sir. We did prepare a contingency plan. I guess I've outlined a number of those options. You've acknowledged that you heard those as part of presentation and the testimony. We'll continue to look at those alternatives using Soyuz as well as Progress vehicles, and we're also hopeful of an expeditious conclusion that would tell us what occurred on Shuttle *Columbia* that would give us an opportunity to return to flight expeditiously.

All those factors are in play. Those are all part of that contingency plan. I think the specific reference in this case from the testimony you cite, was no way to get back with Shuttle immediately given the safety considerations that we will always ground the fleet under those circumstances.

I fully concur on your opening comments, too, as it pertains to charter revisions, to the extent they are necessary. As I pledged to Chairman Boehlert, we will make those changes in any way that Admiral Gehman feels he has to have in order to guarantee that independence.

I have no difficult whatsoever understanding his requirements for independence, and he has reiterated those, and I intend to comply exactly with that approach.

Chairman MCCAIN. Senator Fitzgerald.

QUESTIONING AN AGING SYSTEM

Senator FITZGERALD. Thank you, Mr. Chairman.

Mr. O'Keefe, the day after the *Columbia* accident, I happened to be having a town hall meeting in Champaign, Illinois, and I asked—there were about two, maybe two-hundred-fifty, people in the room—I asked them whether they thought we should continue with manned exploration of space, and I explained to them that it could cost us billions of dollars and years to make ourselves able to continue going forward in space.

To my surprise, I'd say about four-fifths of the people in that room wanted us to go back and continue human exploration of space. And my state of Illinois has very little in the way of spending that it benefits from. We're not Florida or Texas. And I want the space program to continue.

And I wondered if you had a gut impression at this point—and I know it's early, but it seems to me we can go in one of two directions. We could spend billions of dollars and perhaps years trying to patch up and fix whatever may be wrong with the Shuttle program, but you're basically dealing with a 30-year-old design. My understanding is there are some 1.2 million checks that have to be done by hundreds of people before a single Shuttle flight can take off. It's extraordinarily complex.

My question is, Do we go forward and spend that time and that money reinforcing the Shuttle program, or would we be better off not diverting the resources to reinforcing this 30-year-old Shuttle program, and, instead, try and proceed with a new vehicle and focus all our effort on that?

Mr. O'KEEFE. Well, thank you, Senator. The factors, I think, that lead to the complexity of the Shuttle and the amount of checking that goes on there certainly is driven by the technology, no doubt about that, the number of moving parts on that asset. But I would submit that any asset we have that we would use, for the purpose of a Reusable Launch Vehicle capability to launch, would also require an awful lot of checks, as well, because of the absolute dedication, the unwavering commitment to safety, that we always pursue.

Every time we launch a Shuttle flight, no matter what that asset would be, it would require, I think, a review of all the systems checks. And the ethos that we have within the agency and all that are part of the community is that if there's a single thing that is wrong or that appears to be wrong in the judgment of any individual, there is a process set up to stop the launch.

Two weeks in advance, there's a Flight Readiness Review that runs to ground every issue involved in that. If there's any residual issues all the way up to the moment of launch, we don't do that. I don't envision that changing. Even if we had a system today that was brand new, I think that same ethos would have to dominate, because we really are committed to that objective to minimize the risk. We'll never drive the risk out of it completely. And so I think that same approach would be employed no matter what assets or capabilities.

Having said all that, if the investigation leads us to conclude that there is anything structurally deficient about the continued safe operations of the Shuttle system, we positively will take that as a very strong element of the investigation findings and make judgments accordingly that may lead us in the kinds of directions you're talking about.

In the interim, again, our approach is, as I discussed with Congressman Rohrabacher, we tear down this system about every eight to ten flights, essentially rebuild it as new, and it goes through that 18- to 24-month Orbiter Major Modification Program. And so every time that asset goes up there, it is as safe as we know how to make it.

We'll never drive the risk out entirely, but we're trying to manage it down to the lowest possible level and assure that anything that appears even vaguely awry is beaten to parade rest before we let the flight take off, and during orbit, as well.

Chairman BOEHLERT. The gentleman's time has expired.

Mr. Calvert.

Representative CALVERT. Thank you, Mr. Chairman.

Thank you, Mr. O'Keefe, for attending today. And certainly my sympathies to the family and to the NASA family that's certainly still grieving over this loss.

I think all of us here today share one thing, and that's that we desire an independent assessment, unbiased and with the highest integrity, to ensure that future astronauts, NASA, this Congress, and the country have confidence in its ultimate result. Certainly you're off to a good start, and I certainly appreciate that, and I know that we do and the country does. But as Chairman Boehlert has indicated, it may be necessary that changes be made to make sure we maintain that confidence. And I'm thankful that you're open-minded to that.

It's reported that we have a certain amount of money appropriated, I believe about $50 million, for NASA to complete this investigation. Is that funding adequate to pursue, in your mind, to the levels that we're discussing?

Mr. O'KEEFE. I don't know, sir. As I understand it, that's part of the current appropriations conference deliberations. I've read the same press accounts you have. I have no other knowledge of what you and your colleagues may have in mind for that Omnibus Appropriations bill provision, and so I can't make an assessment of that. And I don't know what the cost of this will be, other than to say that whatever it costs, that's what we ought to spend in order to be sure that we reach the answers to what caused this accident.

Representative CALVERT. Obviously, NASA is not the only agency that's going to be involved in this investigation. Are you receiving cooperation from other agencies, full cooperation?

Mr. O'KEEFE. Yes, sir. It's overwhelming. There is no hesitancy, there is no confusion of how that process works. I've just been amazed to see how forward-leaning 20 different Federal agencies, state and local law enforcement officials from Texas and Louisiana, have been in helping us work through what is a real nightmarish circumstance in a way that's professional, aboveboard, and fully cooperative. No hesitation on that point at all.

Representative CALVERT. And that also would apply to the contractors that are involved in this program?

Mr. O'KEEFE. Yes, sir, absolutely.

Representative CALVERT. I know I've read the press quotes. You mentioned, just as of yesterday, that you had no favorite theories of what occurred, and I understand that. However, as we move forward in this Congress, I guess that what Mr. Rohrabacher and others have indicated, do you believe, because of the age of the Shuttle, there is any systemic problem that may be there? And what's our alternatives if, in fact, that's the case?

Mr. O'KEEFE. Again, none that I'm aware of. And, again, we go through an exhaustive process to assure that that the safety-of-flight operation is adhered each and every time. This is not a one-shot deal. It's every—every time it comes back, the orbiters return, we do a careful inspection, we go through a very exhaustive review of everything, and we do not roll it out immediately. There's an orbiter processing facility effort that goes on for the better part of three months as we move it through an exhaustive examination. And then when it gets out to the launch pad, typically it's there for the better part of 30 days in order to make sure that every single thing checks out.

So if there's something systemically wrong, we will be guided by the Gehman board's view of that and we'll correct it. But based on our assessment right now and everything we've done, it sure doesn't look like a systemic failure. But if it is, we positively will correct that before we launch ever again.

Representative CALVERT. Thank you.

Mr. O'KEEFE. Thank you, sir.

Representative CALVERT. Thank you, Mr. Chairman.

Chairman MCCAIN. Senator Dorgan.

SUGGESTING A PRESIDENTIAL COMMISSION

Senator DORGAN. Mr. O'Keefe, thank you for being here today. I think most of us feel that a nation that doesn't explore is a nation that's standing still, and this space program must continue exploring the frontiers of space.

I want to ask you a question, and I don't want you to think the origin of my question poses any distrust for you or the men and women of NASA. I have great admiration for your leadership and also for the men and women of NASA. But as we attempt to find out what happened with this tragedy, it seems to me that in almost any circumstance of this type, an agency can't very effectively investigate itself. I feel there ought to be a Presidential Commission empaneled. I would ask the question, Have you had a chance to visit with President Bush about the prospect of that? And can it be done enveloping reconfiguring the kind of commission that you have now created?

Mr. O'KEEFE. Yes, sir. Oh, yes, indeed, we have visited on the question, to the President as well as the Vice President and all the senior staff on the issue. And I guess the approach that—history is a guide in these cases.

In the *Challenger* accident, it was five days after the accident that President Reagan announced the intent to appoint a commission. It was the better part of 10 days to two weeks before they assembled. It was probably the better end of three to four weeks before they were fully prepared to engage and really start taking testimony and doing the things that were necessary. And they still, nonetheless, produced a set of findings and recommendations by June of 1986. So roughly six months after the accident they were able to reach some conclusions.

In this circumstance, given the development of this contingency plan that we've put together as a lesson learned from *Challenger,* and there was an awful lot that we learned out of that event, that really informed us about how we ought to go ahead and look at ourselves and how we do business. And what it called for as part of that contingency plan was to identify, by positions, the kinds of people that ought to be activated, who are non-NASA individuals and experts, and mobilize them right away.

And so as a consequence, what we defaulted in favor of in this case was speed. We had an opportunity then to have all the members except one, who was a NASA center director of a non-space-flight center, who has no involvement with space flight at all, who was appointed to that particular board. Everybody else is removed from it, and we're moving ahead in that regard as independent as we can possibly make that.

Senator DORGAN. Can I make the point that I think you did exactly the right thing, because you don't want time to elapse. You took action and did the right thing. I'm asking, I guess, as we go along, whether a presidential commission can now envelope, reconfigure the commission that you have started.

I really do think that a year from now, three, five years from now, the question people will ask is, Could NASA really have investigated itself? Again, I don't say that with any distrust at all. I think you've got a great organization. But I really do hope, as we go along here, we're finding a way to perhaps have a Presidential Commission. We don't want to duplicate different investigations, but I think this could be done in the right way and will resolve these questions of independence.

Mr. O'KEEFE. Yes, sir.

Senator DORGAN. So let me wish you well, and please extend, on behalf of all of us in the Congress, our thoughts and prayers to the men and women of NASA.

Mr. O'KEEFE. Thank you, Senator. We're committed to exactly the same objective. We want to find the answers, and we want it to be credible. I mean, there's no question about that at all. So whatever it's going to take in order to do that, that's what we are committed to doing.

And the process, again, is not investigating ourselves. This is an independent group of folks who have no, baggage to carry as it pertains to, NASA biases. Admiral Gehman came from a distinguished naval career that had no involvement whatsoever with NASA, and yet, at the same time, I think he's had a lot of experience, as all the other members did, of better than 50 different investigations into accident situations.

So this is not a group of NASA investigating itself. This is going to be an independent group that's going to reach some conclusions, and we want to make sure that's as credible as we can possibly make it, because that's going to turn on—I think the trust and confidence of the American people depend upon that. Your point is exactly right. I associate myself with that sentiment, as well.

Chairman BOEHLERT. Mr. Lampson.

Representative LAMPSON. I want to thank you, Mr. O'Keefe, for coming to Capitol Hill to testify today. As the member of Congress who represents the Johnson Space Center, I would also like to thank you and your NASA team for the support and encouragement that you've provided to the space center community in Houston during this very difficult time.

I'm somewhat of a reluctant participant in this hearing. Today is the 11th day since the tragic loss of Space Shuttle *Columbia*. There's so much that we do not yet know and perhaps some things that we may never know.

It's my understanding that there were no Congressional hearings on the *Challenger* investigation in 1986 until after the Rogers Commission completed their report four months later. And while I know we're operating under different circumstances, with three astronauts orbiting the Earth in the International Space Station, I do question the merits of having this hearing so soon after the *Columbia* Shuttle accident.

I believe Congress needs to allow the investigation to move forward and to let the accident investigation board members do their work. Hopefully we will complement your efforts and not impede the process.

That being said, I firmly believe that the Administration needs to move forward with a truly independent investigation similar to what President Reagan appointed in 1986 after the *Challenger* accident.

I think NASA made a good first step by revising the board's charter last week, but I still believe, as my colleagues have stated, that NASA's external investigation team is too closely tied to the agency.

As NASA Administrator, the board's charter allows you to appoint the team members, to staff the board with NASA employees, and to receive the final report. In order for this review to have credibility, I believe it needs to have team members who are truly independent and who report to the White House and Congress.

Also, seeing all the cameras and the media presence in this hearing room today begs the question, Where was all this attention to our human space flight programs before February 1st? While I applaud the renewed interest, I regret that it takes the loss of seven fine astronauts for our space program to make the front page of the newspaper or the top story on the evening news.

And while it may seem routine, the work that is being done by NASA in outer space is far from routine. We're doing so many great things in space that benefit us right here on Earth. My hope is that somehow this terrible tragedy will spur the Administration to develop an interest in a real, truly robust space program.

And I'd like to call for a new space race for the 21st century. This space race is not against the old Cold War enemy or an emerging power in the East, but rather our new space race needs to be against ourselves for our own future.

ISS CONTINGENCY PLANNING

And let me ask two things, Mr. O'Keefe. First, a copy of the contingency plan for the International Space Station that you referred to a few minutes ago, could you possibly get that to us within the next week or so? We would appreciate it.

[The information follows:]

Copies of the following documents have been provided to the Committee:

The **Agency Contingency Action Plan (CAP) for Space Flight Operations (SFO)** lays out very specific steps that NASA management/leadership follow when a contingency occurs. The CAP is updated and transmitted to Agency leadership prior to each Space Shuttle launch. The process that I described during my testimony outlines the steps that were taken when Mr. Readdy identified a contingency on February 1, 2000. These steps, including formation of the independent Columbia Accident Investigation Board (CAIB), were taken in accordance with the CAP. (See Tab 1)

The CAP SFO requires that the Director of the Johnson Space Center (JSC) be responsible for preparing a Center plan to cope with any International Space Station Program contingency within JSC jurisdiction. In response to this requirement, JSC maintains an **International Space Station Contingency Action Plan for Lyndon B. Johnson Space Center** (SSP 50190). This plan defines the responsibilities and procedures used by JSC organizations for: (a) taking initial actions in the event of a mission contingency; (b) investigating and reporting mission failures, accidents, or incidents that are under JSC jurisdiction; and (c) supporting investigations as directed by NASA Headquarters. (See Tab 2)

In addition to the above referenced guiding HQ and JSC contingency actions plans, the JSC Space Station Program Office maintains an **International Space Station Program Off-Nominal Situation Plan** (SSP 50562). The purpose of this plan is to ensure that timely decisions and plans are made to preserve the ISS mission plan and mitigate any potential safety risk or impact on research. For example, a specific off-nominal situation (ONS) is "All Shuttle flights delayed greater than 180 days". For this ONS the response is to: (1) reduce operations demands to a level sustainable by Progress and Soyuz vehicles; (2) support operations with European and Japanese launch vehicles when they become available; and (3) evaluate the need for a reduction in the number of crew, or return of all crew. (See Tab 3)

And then let me ask, in 1999, when problems with the experimental X–33 Reusable Launch Vehicle demonstrator made it clear the Space Shuttle would have to be relied on for many more years, perhaps until 2020, the Clinton Administration's OMB sensibly increased the Shuttle upgrades budget significantly. However, in 2001, the Bush Administration's OMB, of which you were deputy director, simultaneously cancelled X–33 program and cut the Space Shuttle safety upgrades budget. How can that possibly have made sense, and can you tell us why you did that?

Mr. O'KEEFE. Again, I'd have to go back and take a look at when NASA cancelled the X–33 program and exactly what was leading to that particular case. But if I can get slide 16, please? The history over the course of time, as I understand it, was a span that you'll see on this particular slide that was for Shuttle funding over the course of that time. The increase that you see occurred, again, as part of the fiscal year '03 budget proposal that we made, and '04, that was just submitted to the Congress last Monday.

[The information follows:]

In 2001, the Space Launch Initiative was just underway, with a goal of replacing the Space Shuttle around 2012. The Space Shuttle safety upgrades program was in its third year, with a goal of developing and certifying the current suite of safety upgrades by 2005 with implementation into the fleet by 2007. This would provide five years of safety benefit prior to the transition from the Shuttle to the new system.

Also in 2001, the X-33 program had reached the end of the time period specified in the cooperative agreement between NASA and Lockheed Martin. When faced with the decision of providing additional funding to continue the X-33 vehicle after the failure of its liquid hydrogen tanks, the Agency decided the X-33 would have to compete for funding with all other reusable space transportation system development efforts under the SLI program. This decision was communicated to Lockheed Martin prior to the completion of the original cooperative agreement, which ended on March 31, 2001. The X-33 proposal was not selected for award under the SLI NRA 8-30 Cycle I competition (May 2001).

In 2001, the electric auxiliary power unit (EAPU) was experiencing technical difficulties, cost growth, and schedule delays. This led NASA, with the endorsement of the NASA Space Flight Advisory Committee (SFAC) and the NASA Advisory Council (NAC), to cancel the project in mid-2001. In the FY 2002 Operating Plan, the Space Shuttle program cancelled or deferred several upgrades because of cost growth or technical immaturity. In the Operating Plan, reviewed by Congress, the funding made available as a result of these actions was then applied to Space Shuttle operations to accommodate operations cost growth. These actions did not affect safety. In September 2002, NASA's Office of Space Flight canceled the supportability upgrade for the Checkout and Launch Control System (CLCS). The decision was based on: unforeseen development difficulties with software, uncertain confidence in meeting schedule, and significant growth in development and projected operations costs, as well as the fact that this upgrade would not have been significantly more capable than the existing Launch Processing System.

Per the latest update to NASA's Integrated Space Transportation Plan, which extends the Space Shuttle's operational life to the middle of next decade, the Administration's FY 2003 budget amendment increased out-year funding for the Space Shuttle program. This increase provides for an additional flight in support of the ISS and an increase in funding for upgrading the Space Shuttle system of approximately $660 million for the FY 2004-2008 timeframe, through a Shuttle Service Life Extension Program (SLEP).

So my reading of the data and the information is that there's an awful lot that contributed to this particular change in funding profile over this span of time, but it was primarily driven by a concurrent, I think, focus on safety improvements and kind of concentrating on all of the factors that would lead to safe-flight operations, and, concurrently, efficiencies that drove down the cost of guaranteeing those particular safe-flight operations through the '90's.

And the most significant increase that's occurred is part of the fiscal year '03 budget amendment the President submitted last November, and the fiscal year '04 budget was submitted last Monday. So those are the primary increases that I've been able to examine, but I'd certainly be prepared to submit all that for the record for your consideration, sir.

Representative LAMPSON. Thank you.

Chairman MCCAIN. Senator Allen.

Mr. O'KEEFE. One other comment, if I could, Senator, is just to reiterate again that the Gehman Commission will report to all of us. He's going to report to the President, to Congress, to all the American people as soon as they reach findings. I have no intention whatsoever of putting any value added to their findings. As soon as the ink is dry, it will be released by Hal Gehman. There is no other approach that I can think of that would be a more appropriate way to handle this so we can move on with finding what the solution is to the problem, get the answers to it, and make the corrections necessary to get back to flying safety.

Representative LAMPSON. Thank you.

Chairman MCCAIN. Senator Allen.

ROLE OF AUTOMATION AND ROBOTICS

Senator ALLEN. Thank you, Mr. Chairman. Thank you, Mr. O'Keefe, for being here.

I want to associate myself with some of the thoughts and philosophy stated in the beginning by our chairman, Senator McCain. And I want to focus on the long-term goals of NASA, broader goals.

If anything good can come out of this tragedy, I think it would be the reinvigorated focus on the mission, primary mission, of NASA, which ought to be scientific research that has benefit for people here on Earth. And I think such sensible strategic planning would be a salutary goal and part of the legacy of the tragic loss of these brave men and women. And I know that of paramount concern to you and all the people in NASA is safety, safety for humans primarily.

Previously, before this tragedy, I know you're on record as supporting refurbished or upgraded Shuttles so they can remain operational for the next 10 to 20 years. I think, in examining the broader goals of NASA, it would be helpful if we'd have some consideration of what is going to be the next orbiter. There are so many questions that we have to determine, and this is just the beginning of this examination. Once we get into our committees in the House and Senate, we'll get in greater detail.

But my question is specific on automation and robotics, and how can robotics and automation and advances in technology, how can that make it safer? It is less costly, but it's also safer for human life. And so is NASA considering an entirely new space plane orbiter or downsizing the manned space flight? Depending on which option is chosen, how will that shape our efforts, our efforts also as the $30 billion, of course, that we've already invested in this space station, the International Space Station, as an investment? But where are we in embracing some of these advancements in automation and robotics? And in the strategic planning, will it effect the continued dangerously underfunding of aeronautics, which I think have tangible benefits to us militarily as well as in the commercial markets?

So I'd like your thoughts on these key paths that we need to go down and decide which ones we're going to go down in the future.

Mr. O'KEEFE. Thank you, Senator. It is, in my judgment, not an issue of either/or, robotics or human space flight. It's how do you do it compatibly? How do you find the appropriate role for robotic capabilities that set, in advance, the kind of knowledge base that you need in order to then support, when necessary, and in circumstances where human intervention and human involvement then becomes very critical.

Again, the Hubble Telescope is the classic example. It's a marvelous piece of machinery that didn't work, and the only way it could be adjusted was to have human involvement in order to make those adjustments on each of the respective servicing missions that have gone on. And now it is rewriting the astronomy books. It is a classic example of how that compatibility between robotics and the use of human space flight intervention, when necessary, can advance the knowledge base dramatically.

But we have to really focus on the risk management side of this and assure that we always use those robotic capabilities, I think, as you've suggested, as a way to fully beat down any of the manageable risk that we see before involving a human space flight capability for that reason, as well as being careful about when you utilize the human involvement dimension to this. That's part of the reason, and a lot of the reason, why the Mars program that we're pursuing for the Mars landers that are planned for later this year and arriving in January of '04 is to advance that knowledge base, understanding fully what's going on in order to then fully support what could be, down the road, a human—a mission that could support that case, if deemed appropriate, necessary, and supported by the research and the science opportunities that could be yielded.

So the strategy you've talked about and the approach that you're alluding to is precisely the direction we're trying to develop now, and have been for some time, as a means to complement those capabilities and always use the robotic capacity up front as the means to inform those judgments.

Chairman BOEHLERT. The gentleman's time has expired.

Mr. Lucas.

Representative LUCAS. Thank you, Mr. Chairman.

Mr. Administrator, down at the Smithsonian, they have a piece of your old equipment hanging for all the world to see, the X–15 from the 1960's, which is a symbol of a debate and a decision by the generation ahead of you and I that, in the spirit of satisfying the common need of the United States Congress and the American people for immediate gratification, it was better to strap men and women and equipment on ballistic missiles than it was to focus on creating space planes.

Your comments today—you point out about the potential future for an Orbital Space Plane and the Reusable Launch Vehicles—with reasonable budget and reasonable focus, how far down the road are we talking about before we have functioning replacement systems like that?

Mr. O'KEEFE. Well, the budget before the Congress as part of the amended fiscal year '03 proposal the President made last November, would contemplate a technology demonstrator of the Orbital Space Plane as early as fiscal year '06, flight testing and so forth to occur as soon as next summer that would lead up to that technology demonstrator. Then, from there to developing as we've now completed the essential baseline requirements, if you will, look for competing approaches—not a technology demonstrator, but an operational vehicle—that would accomplish the objectives of both rescue and return capacity as well as transfer to the International Space Station. It would be online, we would hope, as early as the end of this decade, and we're kind of moving in that direction to try to establish that.

This would be a complementary capability to the Space Shuttle and use the Space Shuttle primarily as a cargo capacity, heavy-lift ability, rather than trying to make a vehicle that's all things to all requirements. This would be a crew transfer capability that would be maneuverable, flexible, and responsive to those kinds of circumstances where needed most.

Representative LUCAS. Booster, slash, plane, or a two-stage plane, Administrator?

Mr. O'KEEFE. It is initially planned as a capability mounted atop an Expendable Launch Vehicle. And that technology demonstrator will be that initial capability that we will utilize at that time.

Representative LUCAS. Along that line, since it's obvious that, with that amount of effort required and the need, as you've pointed out so succinctly, to keep the workhorse, the old Shuttle, up and going, could you address for a moment some of the discussion we've had on the committee for some time about the effect on the reduction in the number of people who—full-time employees who support the Shuttle over the last decade—literally, what, one-third less people still making, if not the same number, but even a greater number of safety checks?

Mr. O'KEEFE. Absolutely. I think the history appears to suggest—and, again, we'll be guided a lot by the review that the investigation board will go through in terms of looking at the systemic causes of what may have been there. So their charter is very broad, and their scope is rather extensive. But it would appear as though that the—exactly as you've suggested, the history is that while cost reductions and efficiencies were gained over the course of that period, as previously described on a slide, there were also improvements in the safety margins as well as the reduction of incidents prior to launch, on-orbit incidents, you name it, there were—all the trends were moving in a direction that proved or demonstrated greater efficiency in addition to slide 18, if you will, that would prove the capabilities, I think, that have significantly improved over the span, both decreasing incidents and increasing efficiencies.

But, again, all that is, is based on the data and the information we see over this particular trend line. We're going to be guided by what the systemic causes are that the investigation board may come back and look at for this information and say that may or may not have been a contributing factor to it. And we'll be guided by their view.

Representative LUCAS. Thank you, Administrator.

Mr. O'KEEFE. Thank you, sir.

Chairman MCCAIN. Senator Boxer.

Senator BOXER. Thank you very much.

Mr. O'Keefe, I want to join my colleagues in sending my condolences to the families and also my feelings of condolence, as well, to NASA. In California, we're the birthplace of the Shuttle program. We hold a very special place in our heart for the heroes who conduct these flights. And it's in this spirit that I ask my questions.

CREW ESCAPE SYSTEMS

In the year 2000, your safety panel made a very clear recommendation. I ask unanimous consent that I place this page in the record. I trust, without objection, that will be done.

[The information referred to follows:]

Ref: Finding #2

The Presidential Commission on the Shuttle Challenger Accident addressed crew escape in their report and recommended that NASA, "Make all efforts to provide a crew escape system...." NASA responded by initiating crew escape studies. Phase I was intended to provide a minimum system prior to return to flight. Phase II was not tied to the return to flight schedule and was intended to provide an automated escape system at a later date uncompromised by the tight return to flight schedule. The Phase II study concluded that an automated escape system was feasible for certain flight regimes and recommended further trade and design studies and a focused development program.

Over the lifetime of the Space Shuttle, a reliable post-launch crew escape system will provide the largest potential improvement in crew safety. NASA has completed or has underway a number of studies that also suggest such a system is feasible. The time is past due for the implementation of a more capable crew escape system.

aerospace safety
advisory panel
annual report for 2000

Senator BOXER. This is a quote, "The Presidential Commission on the Shuttle *Challenger* Accident addressed crew escape in their report and recommended that NASA make all efforts to provide a crew escape system. NASA responded by initiating crew escape studies." This is in this safety panel. Then it says, "Over the lifetime of the Space Shuttle, the reliable post-launch crew escape system will provide the largest potential improvement in crew safety. NASA has completed or has underway a number of studies that also suggest such a system is feasible." And then they say, "The time is past due for the implementation of a more capable crew escape system."

Now, Mr. O'Keefe, after that report was filed, members of the safety panel were fired. And I ask unanimous consent to put in the record the New York Times story entitled *NASA Dismissed Advisors Who Warned About Safety.*

Mr. Chairman, will you put that in the record for me? Mr. Chairman? Mr. Chairman?

Chairman BOEHLERT. Without objection.

Senator BOXER. Thank you.

[The information referred to follows:]

The New York Times

February 3, 2003, Monday, Late Edition - Final

SECTION: Section A; Page 1; Column 1; National Desk

LENGTH: 1414 words

HEADLINE: LOSS OF THE SHUTTLE: SAFETY;
NASA Dismissed Advisers Who Warned About Safety

BYLINE: By WILLIAM J. BROAD and CARL HULSE

BODY:
When an expert NASA panel warned last year that safety troubles loomed for the fleet of shuttles if the agency's budget was not increased, NASA removed five of the panel's nine members and two of its consultants. Some of them now say the agency was trying to suppress their criticisms.

A sixth member, a retired three-star admiral, Bernard M. Kauderer, was so upset at the firings that he quit the group, NASA's Aerospace Safety Advisory Panel, a group of industry and academic experts charged with monitoring safety at the space agency.

NASA said it changed the charter of the group so that new members, younger and more skilled, could be added. "It had nothing to do with shooting the messenger," said Sonja Alexander, a spokeswoman at NASA headquarters in Washington.

Members of Congress who heard testimony from the panel last spring said yesterday that they would re-examine whether budget constraints had undermined safety, but several said they doubted it. The Bush administration said that it would propose a $470 million increase in NASA spending today, and that the increase was planned before the Columbia's destruction.

Dr. Seymour C. Himmel, who was fired from the advisory panel, said yesterday that "we were telling it like it was and were disagreeing with some of the agency's actions."

The eight departed panel members and consultants had long experience with the shuttles' systems and their troubles. In interviews yesterday, some said NASA had developed an institutional myopia about the panel's warnings, advice and observations, however pointed.

The panel's most recent report, which came out last March and included analyses by the six departed members, warned that work on long-term shuttle safety "had deteriorated." Tight budgets, it said, were forcing an emphasis on short-term planning and adding to a backlog of planned improvements. The report called for sweeping change.

"I have never been as worried for space shuttle safety as I am right now," Dr. Richard D. Blomberg, the panel's chairman, told Congress in April. "All of my instincts suggest that the current approach is planting the seeds for future danger."

His worry, he continued, "is not for the present flight or the next or perhaps the one after that." He added, "One of the roots of my concern is that nobody will know for sure when the safety margin has been eroded too far." He could not be reached for comment yesterday.

Leading members of Congressional committees with oversight of the space program promised yesterday that they would investigate whether the budget policies of the administration and Congress were a factor in the loss of the shuttle.

"A large part of our inquiry will be examining what policies contributed to the loss of the Columbia and what policies should follow the tragedy," said Representative Sherwood Boehlert, a New York Republican who heads the House Science Committee. But he said he believed the agency had been adequately funded. "Have we done the right things?" he said. "I think the answer is yes."

The breakup of the Columbia as it began re-entry Saturday morning has put renewed focus on a series of government and independent reports that questioned the fitness of the aging shuttle fleet, the impact of scarce federal money, competing priorities and programs at NASA and a changing work force.

As recently as last week, the General Accounting Office said that the space agency was continuing to be challenged by shortages of trained staff members. Over the years other panels have issued similar reports. For example, a NASA committee reported in 2000 that more money and staff members were needed to support operations critical to shuttle safety.

Some lawmakers also contend that the shuttle program has been shortchanged in recent years while the International Space Station now under construction experienced cost overruns. They said budget problems prevented NASA from initiating safety upgrades in the shuttle.

The new NASA administrator, Sean O'Keefe, has been struggling since his appointment to control space station costs.

Senator Kay Bailey Hutchison, Republican of Texas, said she was "very concerned that we were diluting our mission with these budget cuts" though NASA assured her that safety remained the agency's top priority.

Lawmakers said they also wanted to explore whether NASA's efforts to reorganize its management approach played any role in the disaster.

"You always have to strike a balance between management efficiency and safety, and in the days ahead the Senate subcommittee I sit on needs to look at how that balance is being set." said Senator Ron Wyden, Democrat of Oregon, a member of the Subcommittee on Science, Technology and Space.

The Aerospace Safety Advisory Panel was set up after the 1967 Apollo fire that killed three astronauts on the ground. It was authorized to study the safety culture of NASA's programs and report to the agency each year on its findings. The panel has wide access to all of NASA's facilities as well was its armies of managers and technicians.

Dr. John G. Stewart, one of the fired consultants, who specialized in studies of NASA's work force, said he was most upset because the firings came in midcycle as the panel was working hard toward its next report. Dr. Stewart had argued for years that NASA's work force cuts were getting dangerous. Yesterday, he said the warnings were finally beginning to be heard when he was forced off the panel.

He said much of the investigative spadework he had recently done at the Kennedy Space Center in

Florida and the Johnson Space Center in Houston went unused.

"These work force issues are very important," he said.

But Ms. Alexander of NASA said the newly added panel members were skilled and knowledgeable. One of them, Robert B. Sieck, was formerly director of shuttle processing at the Kennedy Space Center and served as launch director for 52 space shuttle launches.

The newly departed members and consultants, Ms. Alexander added, had an average tenure on the panel of 12 years. The panel's charter, she said, was changed in April 2001 to require a rotation of the membership.

"They were forced out in order to refresh the panel with members carrying skill sets applicable to new technologies and ideas" on how to make the shuttle and other NASA projects safer, she said.

Dr. Norris J. Krone, a fired panel member who heads the University Research Foundation at the University of Maryland, which does civil aeronautics research, said he resented the manner in which he and his colleagues were fired.

"It's unusual to terminate people from a high-level group like that in midterm," he said. "We all thought it was ill-advised."

The White House said late yesterday that Mr. O'Keefe, NASA's administrator, would meet the president Monday and then "fully inform" chairmen and ranking members of the committees and subcommittees with jurisdiction over NASA and its budget.

Staff members of the House Science Committee, which delivered the main Congressional report on the 1986 Challenger explosion, were researching records of the Challenger inquiry yesterday. They were also trying to assemble data on the NASA budget to show precisely the history of funding on the shuttle program and shuttle safety.

"We are going to let everything see the light of day," said Congressman Boehlert, who said an initial review could find no evidence that Congress ever denied a NASA request for resources pinned to safety.

Lawmakers and other space experts on Capitol Hill, however, said it was no secret that NASA has had major difficulties.

"NASA has got a lot of problems, there is no question about it," said one senior official. "They have been under a lot of scrutiny because of some high profile screwups and the enormous cost overruns in the space station."

No hearings have yet been set on Columbia, but lawmakers want to move ahead quickly. The House science panel already had a major session scheduled for Feb. 27 on NASA and officials now expect that hearing to be expanded and refocused.

Lawmakers and staff members who will help run the hearings say they intend any House and Senate sessions to complement the NASA and outside inquiries.

"Everyone is going to be working off the same facts but looking at them from different angles," said David Goldston, chief of staff for the Science Committee. "I think that is healthy but it all has to be coordinated."

Senator BOXER. And after that report and after the people were fired, four board members were fired, two consultants were fired, one board member quit because he was upset at the firings. That left you two people. You changed the charter of the panel.

And I ask unanimous consent that the new charter and the old charter be placed in the record.

Chairman BOEHLERT. Without objection.

[The information referred to follows: the old charter, dated April 29, 1999; the new charter, dated May 1, 2001. Please see Appendix 2: Additional Material for the Record for the charter dated May 1, 2003.]

NATIONAL AERONAUTICS AND SPACE ADMINISTRATION

CHARTER OF THE

NASA AEROSPACE SAFETY ADVISORY PANEL

ESTABLISHMENT AND AUTHORITY

The Aerospace Safety Advisory Panel (hereafter called the "Panel") was established as an advisory committee to NASA by section 6 of the NASA Authorization Act of 1968, as amended (P.L. 90-67, codified at 42 U.S.C. 2477). The NASA Administrator, having determined that it is in the public interest in connection with the performance of Agency duties under the law and with the concurrence of the General Services Administration, hereby renews the Panel pursuant to the Federal Advisory Committee Act, 5 U.S.C. App. §§ 1 et seq. This charter replaces NASA Policy Charter 1156.14N, which is hereby canceled.

PURPOSE AND DUTIES

1. The Panel shall review safety studies and operations plans referred to it and shall make reports thereon, shall advise the Administrator with respect to the hazards of proposed or existing facilities and proposed operations and with respect to the adequacy of proposed or existing safety standards and shall perform such other duties as the Administrator may request. In the pursuit of these duties, the Panel will review, evaluate, and advise on those program activities, systems, procedures, and management policies that contribute to risk and provide identification and assessment for management. Priority will be given to those programs that involve the safety of human flight.

2. The Panel will draw on the expertise of its members and other sources to provide advice and make recommendations to the Administrator. The Panel will hold meetings and make site visits necessary to meet its responsibilities. The Panel will review reports and recommendations of its subgroups and include them if approved and appropriate in its reports to the Administrator.

3. The Panel shall function solely as an advisory body and will comply fully with the provisions of the Federal Advisory Committee Act.

MEMBERSHIP

1. The Panel consists of a maximum of nine members who are appointed by the Administrator. Appointments are made for a single term of 6 years and reaffirmed annually. Renewals of additional 6-year terms are only made with the approval of the Administrator. In order to provide continuity, no more than one-third of the members of the Panel are appointed every 2 years.

2. The membership of the Panel will, to the extent practical, be balanced fairly in terms of technical expertise, professional perspectives, diversity, and the Panel's functions. Members of the Panel shall not be current NASA employees.

3. A Chair shall be selected by the Panel from its membership. The chair will propose meeting agendas, to be approved by the Executive Director, and will officiate at all meetings of the Panel. The designation of any other officers will be at the Panel's discretion.

4. The Associate Administrator for Safety and Mission Assurance may participate in Panel activities but does not serve as a Panel member.

ADMINISTRATIVE PROVISIONS

1. The Panel will report to the Administrator. NASA responses to recommendations of the Panel will be made by the Administrator.

2. Meetings of the Panel will be held as often as necessary and at least once a year. Meetings will be open to the public, except when the General Counsel and Advisory Committee Management Officer determine that the meeting or a portion of the meeting will be closed to the public in accordance with the Government in the Sunshine Act or that the meeting is not covered by the Federal Advisory Committee Act.

3. The Executive Director will be appointed by the Administrator and will serve as the Designated Federal Official. The Executive Director serves as Executive Secretary and Technical Assistant to the Panel and will administer the affairs of the Panel, including arrangements for reviews and evaluations, ensuring that a written record of all transactions is maintained and submitted to the Panel for approval at each subsequent meeting, and taking those actions necessary to ensure continuity and adequacy of Panel efforts.

4. The Office of Safety and Mission Assurance will provide staff support and operating funds for the Panel and administrative management of its subgroups. The estimated annual operating cost totals approximately $523,000, including 3.0 workyears for staff support.

5. Consultants to the Panel may be appointed for a specific task and a predefined period of time by the NASA Administrator or Associate Deputy Administrator, upon the advice of the Panel Chair.

6. The Panel will be provided with all information required to discharge its advisory responsibilities as they pertain to both NASA and NASA-contractor activities.

7. Members of the Panel will be compensated for their services in accordance with section 6 of the NASA Authorization Act of 1968, as amended (P.L. 90-67, codified at 42 U.S.C. 2477) and will be allowed travel and per diem expenses as authorized by 5 U.S.C. 5701 et seq.

8. The Panel Chair, with the concurrence of the Executive Director, may establish informal subgroups to advise the Panel on particular topics. The members of such subgroups will be members of or consultants to the Panel. Subgroups will generally be temporary in nature, will operate consistent with the guidance of this Charter and the Federal Advisory Committee Act, and will report to the Panel.

DURATION

The Panel shall terminate 2 years from the date of this charter, unless terminated or renewed by action of the NASA Administrator.

Daniel S. Goldin
Administrator

APR 29 1999
Date

National Aeronautics and Space Administration
Charter of the
Aerospace Safety Advisory Panel

1.0 PURPOSE

This charter sets forth the authority for, and the duties, procedures, organization, and support of, the Aerospace Safety Advisory Panel.

2.0 AUTHORITY

2.1 Establishment: The Aerospace Safety Advisory Panel (hereafter called the "Panel") was established as an advisory committee under Section 6 of the National Aeronautics and Space Administration Authorization Act, 1968, as amended (Pub. L. 90-67, codified at 42 U.S.C. 2477). The NASA Administrator, having determined that it is in the public interest in connection with the performance of Agency duties under the law and with the concurrence of the General Services Administration, hereby renews the Panel pursuant to the Federal Advisory Committee Act, 5 U.S.C. App. §§ 1 et seq.

2.2 Other Statutory Authorities: 5 U.S.C. Appendix, "The Federal Advisory Committee Act" (Pub. L. 92-463). 5 U.S.C. § 3109, Employment of experts and consultants.

2.3 Executive Orders: Executive Order 12024, dated 1 December 1977. Executive Order 12838 (58 FR 8207, 2/12/93), dated 10 February 1993.

3.0 DUTIES

3.1 Statutory Duties: The duties of the Panel are set forth in 42 U.S.C. § 2477, as follows:

"The Panel shall review safety studies and operations plans referred to it and shall make reports thereon, shall advise the Administrator with respect to the hazards of proposed or existing facilities and proposed operations and with respect to the adequacy of proposed or existing safety standards and shall perform such other duties as the Administrator may request." In the pursuit of these duties, the Panel will review, evaluate, and advise on those program activities, systems, procedures, and management policies that contribute to risk and provide identification and assessment for management. Priority will be given to those programs that involve the safety of human flight.

3.2 Specified Duties: In order to perform their tasks, members of the Panel and Panel consultants must become familiar with relevant NASA plans, policies, and guidelines including applicable NASA Policy Directives, and NASA Procedures and Guidelines (NPG); the NASA Strategic Plan; the Strategic Management Handbook, NPG 1000.2; applicable Center Implementation Plans; and NASA's Agency Safety Initiative (ASI) guidelines of Safety for the Public, for Astronauts and Pilots, for Employees, and for High-Value Equipment. Pursuant to carrying out its statutory duties, the Panel will review, evaluate, and advise on those program activities that in its judgment contribute to safety risk and provide identification and assessment of these for the NASA Administrator. Priority will be given to those programs that involve the safety of human flight. As part of its duties, the Panel may examine any element of NASA's activities that it believes involves safety, including the elements listed below--

3.2.1 Public Safety: This element includes those activities which provide protection for the well-being of people and prevention of damage to property not involved in NASA's business, but which may nevertheless be exposed to potential hazards associated with carrying out this business. Public safety activities performed within the Agency include, but are not limited to, such functions as--

1) Determination of public safety criteria.

2) Establishment and control of public safety hazards associated with facility and systems tests and operations.

3) Establishment and implementation, as required, of emergency or catastrophe control plans.

3.2.2 Astronauts and Pilots: This element includes those activities which provide protection for the well-being of NASA's Astronauts and Pilots. Astronaut and Pilot safety activities performed within the Agency include, but are not limited to, such functions as--

1) Determination of flight operations safety criteria.

2) Establishment and control of Astronaut and Pilot safety hazards associated with flight systems tests and operations.

3) Establishment and implementation, as required, of emergency or catastrophe control plans.

4) Establishment and implementation of safety standards and procedures for operation of program support and administrative aircraft.

3.2.3 Industrial Safety for Employees and High-Value Equipment: This element includes those activities which provide protection for the well-being of personnel and prevention of damage to property involved in NASA business and exposed to potential associated with carrying out this business. Industrial safety relates especially to the operation of facilities in the many programs of research, development, manufacture, test, operation, and maintenance. Industrial safety performed within the Agency include, but are not limited to, such functions as--

1) Determination of industrial safety criteria that protect employees and high-value equipment.

2) Establishment and implementation of safety standards and procedures for operation and maintenance of facilities, especially test and hazardous environmental facilities.

3) Development of safety requirements for the design of new facilities.

3.2.4 Systems Safety: This element includes those activities specifically organized to deal with the potential hazards of complex research and development systems that involve highly specialized areas of technology. It places particular emphasis on achieving safe operation of these systems over their life cycles, and it covers major systems for aeronautical and space flight activities, including associated ground-based research, development, manufacturing, and test activities. Systems safety activities performed within the Agency include, but are not limited to, such functions as--

1) Determination of systems safety criteria, including criteria for crew safety.

2) Performance of systems safety analyses.

3) Establishment and implementation of systems safety plans.

3.2.5 Safety Management: This element includes both the program and functional organizations of NASA and its contractors involved in the identification of potential hazards and their elimination or control as set forth in the foregoing description of safety activities. It also includes the management systems for planning, implementing, coordinating, and controlling these activities. These management systems include, but are not limited to, the following:

1) The authorities, responsibilities, and working relationships of the organizations involved in safety activities, and the assessment of their effectiveness.

2) The procedures for insuring the currency and continuity of safety activities, especially systems safety activities which may extend over long periods of time and where management responsibilities are transferred during the life cycles of the systems.

3) The plans and procedures for accident/incident investigations, including those for the followup on corrective actions and the feedback of accident/incident information to other involved or interested organizations.

4) The analysis and dissemination of safety data.

4.0 PROCEDURES

4.1 Panel and NASA Administrator: The Panel will report to and function in an advisory capacity to the NASA Administrator.

4.2 Information Gathering and Reviews: The Panel will be provided with information necessary to discharge its advisory responsibilities which pertains to both NASA and its contractors' safety activities. This information will be made available through the mechanism of appropriate reports and by means of in situ reviews of safety activities at the various NASA and contractor sites, as deemed necessary by the Panel and arranged through the Executive Director for the Administrator. The Panel shall send observers to NASA safety reviews to the maximum extent possible including Prelaunch Assessment Reviews/Joint Prelaunch Assessment Reviews and Flight Readiness Reviews/Joint Flight Readiness Reviews.

5.0 ORGANIZATION

5.1 Membership:

5.1.1 Panel Members: The Panel will consist of a maximum of nine members, who will be appointed by the NASA Administrator. Appointments will be for a term of 6 years. The NASA Administrator may, at his discretion, choose to extend a member's term by no more than 6 years beyond that member's original term limit.

5.1.2 Consultants: The NASA Administrator may appoint consultants at the request of the Panel for a specific task. A consultant must have expert knowledge for the specific task assigned. A consultant shall serve no longer than required for the effort defined by the Panel Chairperson, or 1 year, whichever is lesser.

5.1.3 NASA Membership: Not more than four members of the Panel can be current employees of NASA, nor shall NASA members constitute a majority of the composition of the Panel and its consultants at any given time.

5.1.4 Ex Officio Membership: The NASA Headquarters Associate Administrator for Safety and Mission Assurance shall serve as the sole Ex Officio member of the Panel. In this capacity, this member may participate in the Panel's activities but not as a voting member.

5.1.5 Membership Selection Process: The Executive Director shall maintain a database of candidates for members and consultants to the Panel. The process of adding a member to the Panel is described in "Appointment of New Aerospace Safety Advisory Panel (ASAP) Members," HOWI 3300-Q025.

5.1.6 Member's Expertise: Membership of the Panel shall be balanced fairly in terms of expert knowledge, professional and/or academic background, diversity, and recent professional experience in their area of technical expertise.

5.2 Officers:

5.2.1 Chairperson/Vice Chairperson: The officers of the Panel shall be a Chairperson and a Vice Chairperson who shall be selected by the Panel from their membership to serve for a 2-year term which may be extended with the written approval of the NASA Administrator.

5.2.2 Powers: The Chairperson, or Vice Chairperson in his/her absence, shall have the usual powers of a presiding officer. The Chairperson may organize the Panel as he/she deems appropriate to accomplish their statutory duties.

5.3 Meetings: Regular meetings of the Panel will be held as often as necessary and at least once a year. One meeting each year shall be designated as the Annual Meeting. The purpose of the Annual Meeting is to present the Panel's Annual Report to the Administrator and the Public. The Chairperson will propose the meeting agenda, to be approved by the Executive Director, and will officiate at the Annual Meeting.

5.4 Reports and Records:

5.4.1 Annual Report: The Panel shall submit an annual report to the Administrator. Instructions on this report's preparation may be found in "Prepare Aerospace Safety Advisory Panel (ASAP) Annual Report," HOWI 1156-Q026.

5.4.2 Safety Reviews and Evaluations: The Panel will submit to the Administrator, at his/her request, reports on all safety reviews and evaluations with comments and recommendations.

5.4.3 Congressional Correspondence/Testimony: The Panel may be requested to provide written responses to congressional inquiries and/or testimony before Congress on specific NASA safety issues.

5.4.4 Files/Records: All records and files of the Panel, including agendas, minutes of Panel meetings, studies, analyses, reports, or other data compilations or working papers, made available to or prepared by or for the Panel, will be retained by the Panel.

5.4.5 Minutes of Meetings: Minutes of all meetings of the Panel, established by the Panel, will be kept. Such minutes shall, at a minimum, contain a record of persons present, a description of matters discussed and conclusions reached, and copies of all reports received, issued, or approved by the Panel. The accuracy of all minutes will be certified by the Chairperson (or by the Vice Chairperson in his/her absence).

5.6 Compensation: Members of the Panel and Panel consultants will be compensated for their services in accordance with 42 U.S.C. § 2477 and will be allowed travel and per diem expenses as authorized by 5 U.S.C. § 5701 et seq.

5.7 Avoidance of Conflicts of Interest:

5.6.1 Nongovernmental Members: Nongovernmental members of the Panel and Panel consultants are "Special Government Employees" within the meaning of 18 U.S.C. § 202.

5.6.2 Financial Disclosure: Nongovernmental members of the Panel and Panel consultants will submit an "Executive Branch Confidential Financial Disclosure Report" (Form 450) prior to participating in the activities of the Panel or acting as a Panel consultant.

6.0 SUPPORT

The NASA Headquarters Office of Safety and Mission Assurance will provide a staff, to be comprised of full-time NASA employees, to provide support to the Panel. The members of this staff shall be fully responsive to direction from the Chairperson of the Panel.

6.1 Executive Director: The Executive Director is appointed by the NASA Administrator and serves as the Designated Federal Official. The Executive Director will serve as the Executive Secretary and Technical Assistant to the Panel. The Executive Director shall have the responsibility for agenda approval, administration of the affairs of the Panel, general supervision of all arrangements for safety reviews and evaluations, maintaining a written record of all transactions, Panel meeting minutes, and trip reports.

6.2 Staff Assistant: The Panel's Staff Assistant shall be responsible for general staff support for the Executive Director, the coordination of personnel and legal forms for the Panel, and other duties as assigned.

6.3 Secretary: The Panel's secretary shall be responsible for general administrative support for the Panel members, Staff Assistant, and Executive Director. The secretary will provide correspondence support, travel support, schedule maintenance and support, and support for personnel and payroll forms.

6.4 Budget: The NASA Headquarters Office of Safety and Mission Assurance will provide the budget for operation of the Panel. The estimated annual operating cost totals $555,000, including 3.0 workyears for staff support.

7.0 CHARTER RENEWAL

This Panel shall terminate 2 years from the date of this charter, unless terminated or renewed by the NASA Administrator.

Daniel S. Goldin

Administrator

MAY 1 2001

Date

Senator BOXER. And, in essence, without going through the bureaucratic talk in here, the new charter, Mr. O'Keefe, gives you much more power—the NASA Administrator, not you personally; in this case, you personally—more power to essentially veto who they choose as chair of the panel.

So I put all these pieces together, Mr. Chairman, and I have concern. I see a report that clearly doesn't mince words here that time is past due for the implementation of a more capable crew escape system. I see members being fired. I then see a new charter where now there's less independence of the safety panel.

I want to know how you feel about this array of facts. First of all, do you agree that the time is past due for the implementation of a more capable crew escape system? And if you do, why haven't we seen more done about it? Number two, why do you think those folks were fired? And, number three, would agree, in light of your, I believe, very sincere comments that safety is a priority, that you would go back to the old charter where the panel could choose its own leader and not have the NASA administrator veto it?

Mr. O'KEEFE. Well, thank you, Senator.

On the first issue, as it pertains to crew escape, again there were a series of very important recommendations that came from the Rogers Commission or outgrowths of the post-*Challenger* experience—that changed operational procedures as it pertained to crew escape and capabilities that were recommended therein. And prior to launch, there is a complete safety regime that's in place that didn't exist prior to the *Challenger,* because of their recommendations. It's a very significant change.

Having said that, my understanding is that the analysis that went on a couple or three years ago following that particular set of reports of the options all led to a series of technical modifications to the Shuttle which have increased its weight dramatically, its operations, its maneuverability, and so, therefore, were deemed to be a marginal improvement in safety that could be attained, if at all, and yet dramatically increased weight, which would have compromised the safety of on-orbit capabilities.

Senator BOXER. So you didn't agree with this recommendation of the——

Chairman MCCAIN. And the gentlewoman's time is expired.

Mr. O'KEEFE. No, I——

Senator BOXER. Well, Mr. Chairman, I'm just trying to see——

Chairman MCCAIN. No, I'm sorry, the gentlewoman's time has expired.

Senator BOXER. I know that you're sorry.

Chairman MCCAIN. Go ahead. We'll recognize the next——

Senator BOXER. I know that you're sorry. Thank you.

Mr. O'KEEFE. I'm sorry, Senator.

No, it is—my agreement, notwithstanding or not, I, again, am not fully aware of all of the parameters of it. I'm advised that's what led the folks to conclude two or three years ago.

Having said that, we are going to look at anything that the investigative board comes back with and says, "These are the changes that must be made in order to guarantee safe flight operations." If it contains that particular set of questions, which, by the way, were primarily pertaining to, as I understand it, ascent requirements, not descent capabilities, that, in turn, those kinds of requirements be factored in and that we make the changes appropriate to do so.

To your second point as it pertains to the safety panel board composition, its charter, and so forth, that occurred prior to my tenure. I don't know exactly what the circumstances were, short of the press accounts and the folklore or legend that may have gone into who did what to who when. Nonetheless, I do understand that, in '97, based on a report from the Inspector General at that time, offered as how a cadre of panel members with long-term experience and in-depth NASA knowledge is important. But to be most effective, this group must be routinely infused with the fresh perspective of new, diverse members. So, as a consequence, the Inspector General's position, as I understand it, was acted upon by my predecessor.

Suggestions were made as to the charter to limit the duration of the tenure to two terms, I believe, of six years each. We'll certainly go back and re-examine that. If it's the desire on the part of the panel members to look at a different tenure period of time that they think enhances their wisdom and understanding of the safety issues, I am all ears on that.

The prior chairman introduced himself to me within 30 days of my arrival at NASA as the outgoing chairman. So I don't know how they arrived at who was going to become the chairman and who would be the next chairperson, but the current chair is the individual that was anointed and appointed, I guess by me, but with the concurrence of the board prior to that time. I made no objection to it. And the only individual who is new to the board is one individual who was added to it during the course of my tenure. No one else has been released.

So I've really been trying to look at what the composition of the panel is, and assure its advisory status, that's the strongest we can possibly make it, and changes made prior to that we'll certainly go back and revisit to assure that if they have different views that would enhance or strengthen their position, that's what we want to hear. We want to make sure that safety of operation is adhered to at all times.

Chairman BOEHLERT. Thank you very much.

Mr. Udall.

Chairman MCCAIN. Could I just say, I want to apologize to all members for enforcing the time limits. We do have such a large number of questioners, and our members have been very patient, and I appreciate that.

Chairman BOEHLERT. Mr. Udall.

Representative UDALL. Thank you, Mr. Chairman.

I, too, want to thank Mr. O'Keefe for taking his time to join us today. And I found your testimony insightful, enlightening, and, in fact, quite moving, and I want to thank you for your leadership.

Mr. O'KEEFE. Thank you.

Representative UDALL. I know you're beginning to think you're in an echo chamber, but I did also want to associate myself with the remarks of our Chairman on the House side and the Ranking Member and others, who have urged you to create as independent a commission as possible and that we'll all be well served when those results are announced.

Mr. O'KEEFE. Yes, sir.

Representative UDALL. In my experience in my previous career as an outdoor educator and someone who was very involved in the climbing and mountaineering communities, we found that when we had accidents, that independent entities that had no fiduciary relationship or other relationship with those involved could make quite accurate and objective determinations of what occurred. So I want to lend my voice to those of others here.

Mr. O'KEEFE. Yes, sir.

Representative UDALL. I did also want to acknowledge the tremendous sacrifice and the bravery of our astronauts and send my condolences to the family members and friends of the brave astronauts. We in Colorado have a proud history of involvement with NASA. In fact, Kalpana Chawla was one of the members of the crew, and she was a graduate of the University of Colorado, so we feel that loss very deeply in Colorado.

Mr. Chairman, if I might, I'd like to include in the record an article from the New York Times on Monday, February 10th, that talks about all the tremendous benefits that have been generated by the space program. I know there are some——

Chairman BOEHLERT. Without objection, so ordered.

Representative UDALL. Thank you.

[The information referred to follows:]

The New York Times

February 10, 2003, Monday, Late Edition - Final

SECTION: Section C; Page 4; Column 3; Business/Financial Desk

LENGTH: 1101 words

HEADLINE: New Economy;
Even as it gazes toward the stars, the space program has broad benefits for those rooted to Earth.

BYLINE: By Amy Harmon

BODY:
SINCE the destruction of the space shuttle Columbia nine days ago, squads of men and women have fanned out across eastern Texas toting state-of-the-art global positioning devices to mark the precise location of fallen tiles, mechanical parts and assorted debris. The devices make their calculations by triangulating with satellites that carry atomic clocks and orbit the Earth at an altitude of 12,000 miles.

Besides performing a task that could help piece together the cause of the calamity, the searchers with their sophisticated global positioning equipment evoke an almost poignant contrast between the creaky 1970's-era space technology that lies scattered across the Southwest and the decidedly snazzy 21st-century commercial technology that is the space program's direct descendant.

"It's really cool that the military put all those satellites up there so we can tie into this and get these positions," said Lee Meeks, a sales manager for Leica Geosystems' United States subsidiary in Norcross, Ga., which usually provides mapping services for counties and is supplying equipment and personnel for the recovery of shuttle debris. "We know within one meter of where we are at any given time with our gear."

The National Aeronautics and Space Administration has come under intense criticism for failing to know much of anything with that kind of accuracy, including what happened to the Columbia. Locked into an expensive, if underfinanced, program developed three decades ago, the agency has essentially been retrofitting and reskinning old technology ever since. Now critics from many corners are calling for a reduction in spending on space, particularly for manned missions.

Meanwhile, many of the technologies that the space program has spawned have taken root and flourished in the commercial world. Among other things, the space program gave rise to the 24 orbiting satellites that make up the Global Positioning System, developed by the United States Air Force as a means of knowing exactly where military troops were at any given time.

Automakers use the system to build devices into car dashboards to enable drivers in many cities around the world to see precisely where they are on a street map displayed on the screen. Car makers also use the virtual wind-tunnel technology developed by NASA to test cars for safety and reliability. The vehicles that emerge are in many ways far more technologically sophisticated than the shuttle itself.

No one argues that potential technology spinoffs are reasons in and of themselves to pour money into NASA. But now, when space technology looks so fragile, may be the time to note some of its often overlooked legacies.

"We're benefiting now from what NASA did 20 years ago and 30 years ago," said Jaron Lanier, a virtual-reality pioneer who developed several technologies for NASA that migrated into commercial devices. "If we stop funding the space program, will we still have the supply pipeline of things that are improving life, or will we be one of those places in the world that feels like the past?"

Twenty years ago, NASA gave Mr. Lanier a contract to help design the first virtual-reality glove, which is supposed to replicate the sensation of touch accurately in a computer-generated environment. Mr. Lanier is now chief scientist of the National Tele-Immersion Initiative, an effort to produce a new generation of Internet technology to provide a physically immersive experience for users. The group is sponsored by the National Science Foundation. Before taking that post, Mr. Lanier parlayed his virtual-reality work into a company called VPL Research, later purchased by Sun Microsystems, which sold gloves and head-mounted displays for entertainment systems. He has also been a consultant for the development of surgical applications for the technology.

Another fruit of the space program is satellite television, made possible by geostationary communications satellites that hover 22,000 miles up, some of which were hauled part of the way into space by the shuttle before using their own propulsion systems to complete the journey. Iridium, a company that eventually went bankrupt, launched orbiting satellites only 400 miles up that still enable people to use wireless phones in places like Afghanistan without needing conventional cellular towers on the ground.

Various plastics, environmental technologies like solar cells, biomedical sensors used by emergency medical technicians, sound-enhancing circuit cards in PC's and the very idea of computer chips have roots in the space program.

"The government doesn't patent these things," noted Arthur Molella, director of the Smithsonian Institute's Lemelson Center for the Study of Invention and Innovation. "What the space program or military research programs allow you to do is nurture these technologies, and there's no question that corporate interests benefit from that. The space program itself doesn't seem to be advancing, but the technologies are taking off outside."

The biggest technological legacy of the space program may be reliable electronics. In the early 1960's, when President John F. Kennedy declared a space race with the Russians, electronics were notoriously capricious. The needs of the Apollo program, historians of technology say, speeded up the early stages of the semiconductor industry by perhaps 10 years.

Soon after the first moon landing, though, Congress began to cut the financing for the space program.

The space program still spawns technologies that make their way into everyday life. The Mars Pathfinder mission, which featured a remote rover that took pictures and explored the planet's surface in 1997, sparked a flurry of activity in robotics.

But the space program's budget, in inflation-adjusted dollars, is a small fraction of what it was during the race to the moon, so, to a much larger extent, the agency draws on technology designed by the marketplace and adapts it to its own ends. Critics say that is a backward flow for technology -- that it is the indirect spinoffs of ambitious government projects that provide the greater public benefit. That is why they say the space program should not be curtailed, or confined to a 1970's field of vision, but should once again train its sights on the future.

"The government does things that cannot be done by the marketplace," said Paul Saffo, a director of the Institute for the Future in Menlo Park, Calif. "And one is to capture the imagination, to give us something we can aspire to. It is the single most important thing."

http://www.nytimes.com

GRAPHIC: Photo: Lee Meeks, a sales manager of Leica Geosystems, uses the company's global positioning device to locate debris from the Columbia shuttle in Nacogdoches, Tex. (Librado Romero/The New York Times)

REPLACING THE SPACE SHUTTLE ORBITER

Representative UDALL. There have been debates and discussions and comments that the astronauts were involved in minor science projects while they were orbiting the Earth. And I think if you look into the record, in fact, what's resulted from our space program is truly remarkable, and day in and day out we see the results of those advancements here on Earth.

If I could, I'd like to focus a little bit on the Space Shuttle orbiter and whether we ought to replace it. Have you gotten to the point where you have an opinion in that regard about the replacement of the Space Shuttle orbiter?

Mr. O'KEEFE. Yes, sir. The Integrated Space Transportation Plan we're currently working with was devised over the course of last year and culminating in the November 13th, 2002, amendment that the President submitted to the 2003 budget that the Congress is still deliberating on at this time. It's reinforced in the 2004 budget submission the President made last Monday, which is to look at all the elements of how these particular systems support each other.

I think, for a long time, all the trends seemed to suggest that every one of these were looked at as individual, standalone programs. But there's a great interrelationship between them. And the requirement for Space Shuttle capabilities, both in terms of crew transfer, which is how we typically have rotated the crews aboard the International Space Station, as well as the launch of cargo assets—in other words, all of the new pieces that are being installed on International Space Station to build out that laboratory that can't be duplicated here on Earth—is a capability we've really got to look at in relationship to each other and to consider a crew transfer and rescue return capacity that can be introduced more aggressively than we presently have.

So the combination of both Shuttle and how we maintain its cargo lift capacity for capabilities to continue to not only support, but finish building, the International Space Station, the capability to transfer crew in order to rotate the expedition crews that we've seen now in our—here we are in our third year of permanent presence onboard that system—as well as the Orbital Space Plane that would provide that capability, all three of those dimensions and the Next-Generation Launch Technologies to ultimately replace the cargo capacity is our focus in that amendment, as well as in the present budget before the Congress right now.

Representative UDALL. Mr. Chairman, I don't know where my time is, but what is the status of the orbiter——

Chairman MCCAIN. Your time has expired.

Senator Wyden.

Senator WYDEN. Thank you, Mr. Chairman.

Administrator O'Keefe, when I chaired your confirmation hearings, I found you to be honest and candid, and we're going to need an awful lot of that in the days ahead, and we appreciate your being here.

NASA WORKFORCE LEGISLATION

My first question deals with the huge brain-drain situation at NASA. It seems to me that you all are hemorrhaging talent in key areas, like electrical engineering. And I think this has implications both for the short-term and the long-term.

The February 1st date, for example, on that date, you all were being pushed to, in effect, use more outside contractors and fewer people within the agency, and so some, of course, are saying that when we have a chance to study this, it's going to back "the people."

So I'd like you to comment on the brain-drain problem, both from the short-term and the long-term, and what's being done to address it.

Mr. O'KEEFE. Thank you, Senator.

Indeed, that is a concern that, as we've discussed previously, as well as we've talked about in various hearings, over the course of the last dozen years or so, we've seen a very clear trend in the direction of an aging workforce that are capable, very strong professionals, but it is, nonetheless, a very mature workforce. We've got three times as many scientists and engineers that are over 60 as we have under 30. And so the consequence of that set of decisions made in years gone by of bringing in additional talent at gradations, there's no way to instantly grow longevity as well as experience base.

What we submitted last June to the Congress was a series of legislative initiatives specifically focused on strategic management of human capital, as has been advised by the General Accounting Office. Dave Walker, as the Comptroller General, has consistently talked about this. So we forwarded this series of legislative provisions. They have been sent to the Congress, they're in the appropriate committees of jurisdiction, and are under consideration to try to deal with what those tools would be that we could use for the purpose of not only retaining for the near-term period the kinds of capabilities and talent we have today, but also recruiting talent with some experience base with a variety of walks and backgrounds, as well as bringing in new graduate students and doctoral students who would replace that roughly 60 percent of the workforce that is of scientific and technical background. You're exactly right, it's a concern, and we want to act on it.

Senator WYDEN. I want to ask——

Mr. O'KEEFE. We look for to the Congress' early enactment of all those provisions to move us along that way.

Senator WYDEN. I want to ask one other quick question. I think when we get to the bottom of this, I think we're going to see that we've got to address this issue, and I just pray that this tragic loss hasn't been due to some human error.

MANNED VS. UNMANNED SPACECRAFT

The second question I had deals with manned versus unmanned space flight. I think that manned flights represent the aspirations and hopes of so many Americans, but I will tell you, I personally believe we're going to need to do more in the unmanned area. I think it is going to be an imperative in the days ahead. And I'd like your judgment as to how to make that call.

For example, I'm attracted to the argument that when you're talking about the space station a few hundred miles, you know, up, that wouldn't be as high a priority as really looking to distant worlds. But I'd be curious how you'd go about tackling this question and making the tough calls with respect to manned versus unmanned space flight. I want to see the manned expeditions go forward, but I do think we're going to have to have a bigger role for unmanned expeditions in the days ahead, and I'd like to hear you tell us how you'd go about making those calls.

Mr. O'KEEFE. Sure, thank you, Senator. I think you've hit the nail right on the head.

The strategy we've tried to employ here, again, is not an either/or, but very much a combination of how do you best employ the robotic capabilities that we have to advance our knowledge base and understand what the challenges will be in order to assure the greatest probability of safety of flight operations when and if called upon to engage humans in that science and research set of objectives.

So the approach that we've devised, for example, in the case of the Mars landers that are planned, and explorers that are planned, for later this year, due to arrive there in early '04, is to continue to build that knowledge base understanding the challenges and difficulties we will work with.

And the inhibitors on exploration much beyond where we are today typically are human related, to be sure, but it's partly technology related. The first one is that our limitations on capacity for propulsion, speed, to get anywhere is currently restricted by the same laws of physics we've been living with for 40 years. And so as a result, until we develop a new space propulsion capacity to dramatically reduce the time as well as the capacity to get anywhere, we're going to be really restricted, in terms of the capabilities we have in that regard.

The second is how to assure that humans survive the experience. And as it stands now, the exposure that we see and that we're learning on International Space Station as a consequence of long duration spaceflight are the debilitating effects on human beings of space travel and space exploration. We're looking to conquer those. Part of the budget proposal you have before you as part of the '04 submission that the President just made is an intensive effort to look at human factors. And only then, after we've conquered those kinds of challenges of degradation, of muscle mass, bone mass, radiation effects, all those things, should we venture much beyond where we have the capacity to do today, which is a very important pursuit of science and research aboard station and other objectives.

So the whole strategy here is to lay this out in a way that informs the knowledge base by robotic capabilities, follow along to the extent necessary and when human intervention gives us the opportunity to expand that knowledge base, and make sure they can only do it when there's a safety-of-flight capability that we can assure.

Senator WYDEN. Thank you, Mr. Chairman.

Mr. O'KEEFE. Thank you, Senator. I appreciate it very much.

Chairman BOEHLERT. Mr. Weldon.

Representative WELDON. Thank you, Mr. Chairman.

Administrator O'Keefe, thank you for coming. And I have the highest confidence in your leadership, as I do in Admiral Gehman's leadership, who did an outstanding job in investigating the USS Cole.

THERMAL TILE ADHESIVE

I have a very specific series of questions that you may not be able to answer here, but I would like a thorough response for the record, relative to one aspect of the operations of the Shuttle, and it deals with the tiles.

The tiles are glued to the Shuttle by a special adhesive. That adhesive has, as it's primary component, urea. The urea that's produced is produced around the world, and much of it's for agriculture and industrial purposes. But the specific urea that NASA has used for the glue for the tiles was produced by one plant, and that one plant was in Fort Saskatchewan, Alberta, Canada. And the reason why that plant was selected was because none of the U.S. manufacturers were able to meet the very stringent requirements that NASA had established for the urea, for the glue for the tiles.

About five years ago, that plant was acquired by another Canadian firm that does business in Cuba. And because of that, they were concerned about the implications of Helms-Burton legislation, and so they no longer supplied NASA the urea for the glue for the tiles.

The U.S. manufacturer of the adhesive that used that specific urea was very concerned at the time about finding a new source of urea that would meet the very specific, tough requirements that NASA had for the glue to hold the tiles on. And I would say there are millions of tons of urea consumed in the U.S. every year. But only a very, very small portion of it would be used specifically by NASA for the glue for the tiles. And, as I said before, up until that takeover five years ago, it was from one plant in Canada that had a separate mechanism for producing that urea that U.S. manufacturers did not, or perhaps could not, achieve the same quality standards that NASA required.

So what my concern is, whether or not we found an equally reliable supplier of urea. And, for the record, I'd like you to give us that information relative to the specifics of NASA specifications.

Thank you.

Mr. O'KEEFE. Yes, sir. No, I don't know. I really am not aware of the nuances there, but I positively will provide that for the record.

Representative WELDON. Thank you.

[The information follows:]

RTV (Room Temperature Vulcanizer) 560 has been the only tile adhesive used in the Space Shuttle program for bonding the insulating tiles to the Orbiter during the life of the program. The formulation of this adhesive has remained the same since the beginning of the Shuttle program and has never contained Urea.

Since the late 1970s, GE Silicones in Waterford, New York, is and has been, NASA's <u>only</u> qualified manufacturer of the RTV.

RTV560 is a two-part system consisting of the base, RTV560, and the catalyst, Dibutyl Tin Dilaurate or DBT. Upon mixing at the correct ratios, the two materials react and cure in a condensation reaction to form a rubber.

As supplied by GE Silicones, RTV560 does not contain Urea. The primary ingredients in RTV560 are Dimethyldiphenylsiloxane (phenyl polymer), Red Iron Oxide (filler/heat stabilizer), Diatomaceous Earth (filler), EthylSilicate 40 (crosslinker) and water. The catalyst is DBT.

Mr. O'KEEFE. Thank you, sir.

Representative WELDON. Thank you, Mr. Chairman.

Senator BROWNBACK. Senator Snowe.

Senator SNOWE. Thank you, Mr. Chairman. And welcome, Mr. O'Keefe. I know this a very trying time for you and the NASA family and most certainly the families of the astronauts. And it just reminds us how fortunate we are as a nation to have been blessed with men and women like these astronauts who are willing to take risks for this country.

DEBRIS ASSESSMENT AND NEED FOR IMAGERY

I'm trying to get at the picture of how NASA approaches certain decisions—what is minimized, what is discounted. We know that for 12 days, from the time that you all learned of the debris that hit the Shuttle and then the Shuttle was scheduled to land, no action was taken other than doing some computer model simulations to predict damage and to rely on past experiences where Shuttles had returned safely, even though there had been several Stanford studies in 1990 and 1994 that had already warned of some potential damage that a single piece of debris could have had on the tiles.

Could you tell me as to why no request was made for military telescope imaging? We know that a camera was not working at the time of orbit that really could have shown the damage that was done on the underside of the Shuttle. Why wasn't that requested at some point in time during the flight to do a greater examination of this type of damage, rather than relying on computer modeling when you really didn't know what had happened, rather than doing the modeling on something that you knew had happened?

Mr. O'KEEFE. Thank you, Senator.

The investigative process, and certainly the Gehman Board, if they come to find we should have done something else, positively we'll be, you know, guided by that particular finding. Nonetheless, the approach that was taken here is, this is a piece of foam material that was about a foot and a half by six inches of which there have been incidents like this before. And, as I mentioned earlier, there are cases where after the flight, there's a full examination of every square inch, every single element of the orbiter when it comes back, to see what the damage effect was. It was determined, in previous cases of comparable circumstance, not to have been a safety-of-flight consideration.

Again, the circumstances here were, it came off of the external tank as the entire Shuttle orbiter system was traveling at 3600 miles an hour. The piece came off, dropped roughly 40 feet at a rate of something like 50 miles an hour, so it's the functional equivalent, as one astronaut described to me, of a Styrofoam cooler blowing off of a pickup truck ahead of you on a highway. And every incident we'd seen before that, every model we ran, every analysis that had been done on every prior case demonstrated no significant damage in that circumstance.

Of the 4,000 sensors aboard the Shuttle orbiter, none of them indicated any anomalies during that 16-day flight. And given the wide variation of heat of several hundred degrees that was experienced 16 times a day, if there was any penetration, any damage that could have been evident, the assumption was those sensors would have picked it up.

Nonetheless, if the Gehman Board finds that we really erred by not examining this in yet another direction, based on all the historical evidence, we positively will run that finding to ground and make corrective actions as necessary.

Senator SNOWE. But wasn't this piece of debris the largest documented piece ever to hit the Shuttle?

Mr. O'KEEFE. Not to my knowledge, but I will correct that for the record if that proves to be in error. I don't know whether that's true or not, but I certainly will provide that for the record.

[The information follows:]

The size of the debris that is believed to have originated in the External Tank bi-pod area and hit the shuttle has not been determined definitively. Therefore, we cannot confirm that this piece of debris as the largest piece to ever hit the Shuttle.

The analysis NASA conducted during the STS-107 mission to determine if the debris could cause damage to the orbiter assumed a total size of the bipod ramp (a piece of foam 20 inches long, 16 inches thick, and 6 inches wide, weighing 2.67 lbs). The analysis used velocities ranging from 550 fps to 945 fps, much higher than the predicted rate of the debris to predict potential damage.

Senator SNOWE. I guess——

Senator BROWNBACK. The time of the senator is up. I'm sorry.

Senator SNOWE. Thank you.

Chairman BOEHLERT. Mr. Wu.

Representative WU. Thank you for being with us during a very difficult time, Mr. O'Keefe.

During my colleagues' questions, I took the liberty of drawing up a little diagram to illustrate my inquiry to you. It's not a PowerPoint presentation; it's just felt tip pen on a piece of paper. Across the bottom here, cuts in your budget. And going up, risk. And the red line is the typical hockey puck kind of curve that some of us in high tech like to see in financial returns, but we don't like to see in this kind of context.

And earlier, I heard you say that you are pounding out as much of the risk as possible before each and every Shuttle launch. But we also have a history of delayed improvements, perhaps delayed in future generations of crafts which may be safer. And I am concerned that the tragic loss of seven astronauts tells us that we are somewhere out on this leg of the curve and not somewhere here, you know, in the flatter portion.

It's our job to try to set policies which maintain reasonable safety, a job which we share with you. You are a very good team player. You should be. But in response to specific congressional inquiry, I think that you are free to answer those inquiries.

And I want to make this a standing congressional inquiry, if you will, that whatever the optimal budget is, as we are adjusting that budget, can you work with us to find that inflection point? I'm concerned that we have gone past that inflection point in risk where the risk has become unacceptably high.

It is always going to be inherently risky to put human beings in space. I'm a strong supporter of human space exploration. But I want to invite you to work with us to find some reasonable point in here where we are not expending exceptional resources, or unnecessarily expending resources, but we are doing everything reasonable to keep humans safe in space.

Mr. O'KEEFE. Sure. No, absolutely, Congressman. I'm delighted to work with you to try to find what that breakpoint is. And, again, my appreciation—slide 18 again, please—is that over the course of time, we've seen a reduction in cost of activities, there has, at the same time, been an improvement in efficiencies as well as the reduction of in-flight anomalies, technical scrubs have dropped by a lot, all of the basic factors that would drive you to conclude that, as your chart suggests, as you reduce resources, you should see an enhancement of risk. If anything, what appears to suggest here is a case where efficiencies have been attained and risk has been reduced.

So the extent there are differences of view about that over the course of this past decade of whether or not that is the contributing factor to it, we really are looking forward to trying to determine how to correct that. And if we've crossed that threshold I think you've so eloquently alluded to, we really ought to figure out exactly where we make those adjustments as necessary.

But the trends are the things that I think we need to analyze here, as well as just the basic theory, that you've advanced, which is a sound one.

Representative WU. Well, this is why I drew it in this way, because if you have effectively reduced cost and reduced risk, you've shifted this curve to the left or to the right, up or down, or diagonally, but the curve is still here——

Mr. O'KEEFE. Yes, sir.

Representative WU.—if you make these assumptions that such a point could be statistically determined. And I just want to invite you, as this curve shifts, as policy shifts, to help us look for this curve. You and I have been in this discussion before——

Mr. O'KEEFE. Yes, sir.

Representative WU.—about the worthiness of human space flight. And I want to remind you of our conversation that Lewis and Clark went west 200 years ago. They got an Appropriation of $2,500. They spent $38,000, and that caused President Jefferson a lot of heartache. But that turned out to be a pretty good deal for America in the long-term.

And I would just encourage you to aggressively ask for what you need and to keep the explorers safe out there.

Senator BROWNBACK. The gentleman's time has expired.

Representative WU. Thank you, Mr. Chairman.

Chairman BOEHLERT. Thank you, Mr. Wu.

Senator Burns.

Senator BURNS. We need you on Appropriations.

(Laughter.)

Senator BURNS. Mr. O'Keefe, I wish we were meeting under different circumstances, but we are not. And my question is a general question, because I was pretty close to the negotiations of the International Space Station and the agreement that we signed with Russia.

And at that time, I asked a question that we really didn't pursue for some reason or other. I think it would help this committee if—as you know, we look at programs and the infrastructure that it takes to carry those programs out. At the time we built the orbiter, was there any estimates of—what every program goes through is, there is a point diminishing returns whenever upgrades are not sufficient to carry out the mission, and I'm wondering if any estimates early on this program were made by engineers of at what point do we come to a point of diminishing returns. And if we could look at that and then—and I know programs change and missions change, and if history tells us anything, we should be looking at those kind of things in order to change the way Congress should be shouldering its responsibility.

Mr. O'KEEFE. Yes, sir.

Senator BURNS. And I would just ask if there were—any research could be done in your records of when do we reach that point, did we reach that point, and what was—and as programs change, what is being dictated in the future if this equipment is going to be asked to do things maybe it was never intended to do.

I'm not going to go over the past, because I've been intimately involved with it. And no other program stimulates the curiosity or the interest in our sciences and our mathematics in our schools like this particular agency of the United States Government. And so I deem it very, very important.

But if we could have a history and see the things that we can do, and then you do what you do best, we may have to call on our older end of the engineers, so to speak, to make those determinations, but I think it would help us a lot if we could reach back there and look at history, take a look at what happened, and then make some decisions to enable you. We don't want to see this happen again, but we know that this will happen. Accidents will happen, especially in the area of going into the unknown.

And I thank you for being here today and some explanations we've reached today. I'm looking for history, something that we base policy on into the future, upon your recommendations.

Mr. O'KEEFE. Well, thank you, Senator. I'd be delighted to provide that. We'll go through that consideration. There is no question that as it pertains to current flight operations, and I want to reiterate, we have a culture that is just obsessing over not letting anything go until it's all exactly right. If the investigation board found that systemically we have failed in that quest, that's precisely what we'll be guided by, as well.

But your point is very well taken. I think we've got to really be thinking seriously about where is that stage where we really make those decisions, and I think we'll provide that, for sure. I'll work through that analysis and provide it for the record, as well.

[The information follows:]

Since the inception of the National Aeronautics and Space Administration, the finest minds in science and engineering have been engaged in the most challenging endeavors of exploration known to mankind. And with a 45 year history, NASA's human space flight program has faced serious adversity with the *Apollo 1* fire, the *Challenger* accident, and now the loss of *Columbia* and her crew. In each case, NASA determined the cause, fixed the problem, and moved forward. Various studies and reports have been written describing the changes that occurred within the NASA program to build on the lessons learned such as changes in program management, changes in technical procedures and changes in risk analysis. Likewise, over the past 17 years the Space Shuttle program has seen technology evolution (glass cockpit) and safety upgrades like those incorporated into the Space Shuttle Main Engines. We learn from history and experience and build upon past knowledge.

In November 2002, NASA finalized the new Integrated Space Transportation Plan (ISTP). The new plan calls for the Space Shuttle to support the ISS through at least 2015. ISTP lays out a plan for our key space transportation capabilities over the next several decades. Consistent with the ISTP, the FY 2004 budget assumes the Space Shuttle will continue to be flown and maintained at least into the middle of the next decade as the Nation's primary vehicle for human access to space. We believe the revised ISTP is a good plan, but we are committed to re-examining it if necessary in light of future investigation findings on the Columbia accident.

The Service Life Extension Program (previously Shuttle upgrades) is an essential element of the integrated NASA space transportation strategy. The Space Shuttle investment reflects a near-term integrated investment to assure that the Space Shuttle can safely support the assembly and sustained operation of the International Space Station (ISS) well into the next decade. The Space Shuttle's Service Life Extension Program addresses the critical requirements for the Space Shuttle to safely and effectively perform its mission until a replacement is available. This effort focuses in seven areas: safety, sustainability, infrastructure, resources, operations, performance, and integration.

Senator BURNS. Thank you for your leadership, and I appreciate your cooperation.

Chairman BOEHLERT. The time of the Senator has expired.

Mr. O'KEEFE. Thank you, Senator. I appreciate it very much.

Chairman BOEHLERT. Thank you very much. We're going to take a brief five-minute break. Five minutes only. And then we're right back. And when we come back, Mr. Nethercutt starts the questioning.

[Recess.]

Chairman BOEHLERT. The Committee will resume.

The Chair recognizes Mr. Nethercutt.

Representative NETHERCUTT. Thank you, Mr. Chairman.

Mr. O'Keefe, I want to welcome you, sir. Over here.

(Laughter.)

Representative NETHERCUTT. I know, I moved.

I appreciate your being here, and I appreciate the sensitivity with which you and the entire NASA team reacted to this terrible tragedy. I certainly was touched by Senator Stevens' remarks and agree with him with respect to your integrity and your qualifications.

Mr. O'KEEFE. Thank you, Congressman.

Representative NETHERCUTT. Thank you.

The crew that we lost touched my Eastern Washington District. Michael Anderson was a proud product of our community, and Ron Dittemore certainly is, too. And so it touched our community very deeply. But in that respect, we're respectful of all that they have done and, in the case of the NASA team, will continue to do.

This was a science-driven crew. They spent 16 days in space and were 16 minutes from landing. And in the process, with the space research double module, we're doing tremendous numbers of experiments, as I understand it. And with the loss of the *Columbia*, the question comes, what data might we have been able to collect with respect to their 16 days of scientific research efforts? And maybe that's my question, basically, is what were we able to retain and preserve with respect to their scientific research legacy?

Mr. O'KEEFE. Yes, sir. No question, it was an extraordinary mission. It was intensively science focused. You've characterized it exactly right. Over 16 days, a lot of the data and returns from many of those experiments were relayed back, and so the scientific community has the benefit of that information. But, to be sure, the physical laboratory as well as the physical experimentation that was aboard STS–107 is lost for all eternity. There's no question there.

But let me provide for you for the record a rundown of the kind of data and information we have gotten back, categorized by the kind of areas. But it was a phenomenal trove of information that I think will yet prove to be very enlightening information as research continues on a range of biomedical as well as physical sciences research and material research activities in the future.

[The information follows:]

NASA scientists are continuing to assess the status of the data received by the experiments onboard Space Shuttle Columbia (STS-107) during its final mission.

Columbia carried more than 80 experiments including science, commercial and student experiments, on a 16-day mission devoted to research, entrepreneurship and education. To date, 27 experiments have reported receiving some data from the mission. For the Biological and Physical Research (BPR) experiments, it is estimated that for those experiments that received downlinked data during the mission, anywhere between 50-90 percent of the data was acquired. Most of these experiments were in the physical science disciplines of combustion research, material sciences and fluid physics. For most of the life sciences experiments, data and specimens were to be recovered on landing, therefore no data is available. The Biological and Physical Research science project teams report the overall performance of the experimental hardware and equipment employed on the mission was highly successful, with 100 percent operational success being achieved for virtually all of the experiments. This operational success provides a measure of the robustness and capability of conducting high quality research on the Space Shuttle and the dedication and training of the crew.

Shortly after the loss of Columbia, the Associate Administrator for BPR sent a letter to the research community directly involved with STS-107, acknowledging the loss of the crew and Columbia, notifying the 107-research community of our intention to support the accident investigation teams, and reaffirming our commitment to ISS and Space Shuttle research. BPR is committed to working closely with its investigator community. Similar to the Orbiter's data, the payload source data are to be considered impounded for use by the Columbia Accident Investigation Board (CAIB). Investigators are asked to contact BPR if they have data that had been previously down linked and has been analyzed for its original research objective. During the period of the CAIB investigation, BPR is setting up an orderly process to collect, review and ascertain every aspect of the payloads, their operations and condition during re-entry. From the assessment, BPR will work with Principal Investigators to see if there is any more data that can be collected in ground facilities to augment what they learned during the STS-107 flight.

Representative NETHERCUTT. Let me ask you if there is any support that NASA will offer to the principal investigators who lost scientific capability as well. Have you been able to assess that yet or make any judgments about the principal investigators and what losses might have been sustained as the *Columbia* was lost?

Mr. O'KEEFE. Yes, sir. There are a number of folks who had based a lot of experimentation, their entire dissertations were riding on this, so years of research activity has really been set back dramatically as a result of that. That's inconsequential, though, by comparison to the loss of lives, to be sure, but it is something we need to be extremely mindful of.

There was to have been a get together this past week, I think, with the biological and physical research components of our agency with all the principal investigators that had a stake, if you will, in the STS–107 experiment and research regime. We are certainly intent on trying to reconvene that session to find out what may be remedial for their efforts as we work through this, in terms of the kind of information we might look to in future flights. But we will work that. I assure you, that's something that's prominent on our minds, as well.

Representative NETHERCUTT. Is your commitment lessened or diminished at all to scientific research and the value of station and the efforts that were undertaken by this crew?

Mr. O'KEEFE. No one iota. As a matter of fact, the families of the STS–107 crew, the most stoic, courageous people you would ever want to meet, within two hours of this activity were already saying, "You know, you cannot give up on this set of objectives. They dedicated their lives to this. That's what they were committed to doing. You cannot move away from it." It had been an inspirational group, and that, in and of itself, has been sufficient cause in my mind to not step back from our commitments in this regard one inch.

I appreciate it, Congressman. Thank you.

Representative NETHERCUTT. Thank you.

Senator BROWNBACK. The time of the Member has expired.

And if I could ask the people operating the door if you could keep that door closed as much as possible, there is some beautiful singing going on outside, but we don't need it in the room.

Now, I have the only astronaut that's serving currently in the United States Senate, Senator Nelson, from Florida.

Senator NELSON. Thank you, Mr. Chairman.

Mr. O'Keefe, prior to you arriving at NASA, the Space Shuttle budget was whacked by some $1.4 billion. Basically, part of that over a nine-year period, this says. And, by the way, it's not the easiest to find this out, because prior to your arrival back in the early '90's, everything was lumped in together into a human space flight account—the Space Shuttle, the kinds of new technologies, plus the station. But when you break it out, what you find is that the Congress whacked part of it, about $600 million, out of the Space Shuttle, and then NASA itself whacked another $750 million.

Putting those two together, you can see the years. And this is prior to you arriving, in '02. That year, the Congress had added some $45 million, and NASA had whacked $70 million to the Space Shuttle.

So this will be an ongoing dialogue that we will have. But the question is, What is your opinion, prior to your arrival, as you look back, what had happened over that nine-year period? Sometimes the Congress would take the money out or just reduce it. Sometimes NASA would basically reprogram the money and take it out the Space Shuttle and put it elsewhere. Does that compromise safety?

Mr. O'KEEFE. Well, sir, I don't know the audit trail, clearly, as well as you've obviously researched this. But this particular, I think, matches with what you've projected here, which is the funding history. And the convergence of two events—and, again, I've got to really look at this in much greater detail to see the individual year changes that have occurred and so forth—but the trends seem to connote two things.

The first one is that at the same time that efficiencies were being yielded and different ways of going about business that are more risk management and more what I would call quality-assurance-related approaches that raises and improves the risk-management probabilities, at the same time also yielded some cost reductions along the way by not having an intensive group of individuals involved in the activity. And so all the indicators over this same span of time seem to suggest—but, again, we've got to back and really look at this very, very carefully—would seem to suggest that there were improvements in incidents prior to launch, incidents on orbit, all of the trend lines that we use to measure the efficiency and performance of the space flight operations program seem to be moving in that kind of a trend line.

That said, we're going to be guided by what the Gehman Board looks at as systemic causes. If this appears to have been a contributing factor, we will be right back here looking at what those fixes need to be to work on that.

Senator NELSON. And we will carry on a continuing dialogue on this. I can tell you, there are people at NASA and in the astronaut office that feel like that safety has been compromised over the last 10 years as a result of the Space Shuttle budget being raided. And that's something that we've got to be concerned about.

One other item——

Mr. O'KEEFE. We're happy to hear those comments and any other views from anybody inside to external to the agency. It'll come to ground truth and find the answers to what happened in this case, absolutely, Senator.

Senator NELSON. Might you comment on the fact that if that——

Senator BROWNBACK. I'm sorry, the time of the Senator has expired. We're having to stay on very tight time frames.

Chairman BOEHLERT. The Chair recognizes Mr. Weiner.

Representative WEINER. Thank you. Welcome.

I fear in your statement you have articulated, I guess, a strawman that some of my colleagues in their questions have knocked down. And when you said that we ought not turn our backs on exploration and that the research that was done on the Shuttle was valuable in cancer treatment, crop yield, and fire suppression, and dust storms. But it is a fact that all of that research could have, should have, and would have been done on the space station had it been completed. And, in fact, the Shuttle has, more often than not, not been a research vehicle, but a delivery vehicle supporting other platforms for science, whether it be satellites or telescopes or the space station.

One of the many things that made the *Columbia* mission noteworthy was the fact that it was a pure science mission. Apparently only 11 of the last 46 Shuttles have been able to say that. It got to be so frustrating that in the 106th Congress there was actually language put into the Appropriation bill that this Shuttle should contain more research.

You know, Shuttle astronauts, I fear to say, have become, more often than not, very high skilled, often brilliant, undeniably courageous cargo carriers. And to demonstrate this point, I don't have a graphic, but you do, and I'd ask you to put up number 20.

Under something marked "safety indicators" is a chart that said launches more than ever cargo capacity up as much as 100 percent. Cargo capacity being an indicator of safety leads me to the inescapable conclusion that having more cargo means fewer flights, means safer human beings. This should not be how we measure whether someone is safe or not, because, frankly, as we learned within, I guess, 48 hours after this horrible accident, an unmanned vehicle went up and brought cargo to the space station.

And also, on chart number 18, the same chart that has the reduction of in-flight anomalies, monthly mishap frequencies, technical scrubs, brags about the increase in lift capacity to the Space Shuttle. It seems that we're mixing the need to keep people safe, which is something that I think you have articulated several times here today, with this ever-growing notion that the Space Shuttle is the only way should develop or the only way, the only means we should use to carry cargo.

We want science to be done in space. Over and over again, we, in Congress, have been asked the question, because our colleagues put it to us, "Do you want to continue the space station funding?" We all say yes. I say yes. But we have to be careful not to confuse what the Shuttle has been as a science mission. It has been a UPS truck for Space Shuttle supplies. And I'm not sure that if you believe that increasing cargo is a way to make people safer than having a manned cargo carrier is the right way to go at all. And if you'd just address that, particularly chart number 20, if you could.

Mr. O'KEEFE. Sure. Thank you, sir.

Cargo, in the term you've used here, means assembly and science, both. In the last four years, many of the Space Shuttle flights, and I think you've pointed out the history precisely right, that we have dedicated the use of Shuttle for the purpose of bringing up large sections of the International Space Station for on-orbit assembly.

This is an engineering marvel we're building in space. You know, there's no other way to do this. There's no way to launch the completed International Space Station in one fell swoop or one piece, so each of it's been assembled on orbit. And by no means are these UPS truck drivers.

Representative WEINER. No, the question, if you'll just understand, the question is not that. It is if you can bring food, if you can bring clothing, why can't you bring Space Shuttle?

Mr. O'KEEFE. I'm sorry, I didn't get to the answer fast enough. I apologize.

Representative WEINER. I'm sorry, I'm——

Mr. O'KEEFE. The Progress vehicle that went up the Sunday after the accident did, in fact, contain groceries, logistics supplies, those kinds of things. That's not typically what we put aboard Shuttle. There are some of those things that are there, but mostly those are carried by those unmanned autonomous capabilities that are brought in to sustain the typical consumable requirements.

Others are put aboard Shuttle, too, like water and a few other things, but typically what is, is the cargo section includes the components, the modules of the International Space Station that couldn't get there any other way, or the science. And as we see in this particular case of the STS–107 as well as on every one of the Shuttle flights, the science experimentation going up-mass to the International Space Station to bring those scientific experiments to there, there's no other way to do that. There's no way to put them aboard autonomous unmanned vehicles at this juncture that would do anything other than provide basic logistics requirements.

So I get your point. You're exactly right. We're trying to maximize the yield of what can be, as we call it, up-mass to the International Space Station or in any other orbit pattern, but, at the same time, also minimize the risk to the individuals so that really the human involvement is minimized to the point where it's actually necessary.

Senator BROWNBACK. The time has expired. Thank you very much.

We now have the Senator from the host state for the Johnson Space Center, state of Texas, Senator Hutchison.

Senator HUTCHISON. Thank you, Mr. Chairman.

And I will just say, Mr. O'Keefe, I know how devastated you are. I've never seen a sadder face than yours in the last few weeks in all the pictures, and I think that you have handled the immediate aftermath very well, and I appreciate that.

Mr. O'KEEFE. Thank you, Senator.

Senator HUTCHISON. I also want to say I appreciate Mr. Rohrabacher's mention of the Texas residents who never expected anything like this but have been so supportive of NASA throughout East Texas looking for the debris, and consider themselves, sort of, deputies in the investigation, and I'm very proud of my home state, and especially the NASA people and the NASA family. I grew up in the area. I have known the NASA family since the announcement that NASA would come to Johnson, and have known the close-knit nature of that community. And I appreciate all of them, as well.

I want to talk about some of the experiments that have been successful and have made a difference in our lives really, from the National Science Biomedical Research Institute, which is not the old, past successes of space research, but the newer ones. They have developed portable infrared sensors to determine blood and tissue chemistry noninvasively, which could help us in intensive care units and ambulances be able to test people quickly and determine hemorrhaging or other maladies; developed a biosensor for microbes and toxins that has an application in the bioterrorism field for early detection and treatment, could be used by military searching caves in Afghanistan or by weapons inspectors; helped further development of a focused ultrasound system for hemorrhage control and for destroying unwanted tissues or tumors that could one day allow bloodless surgery.

Right there on the *Columbia*, they dealt—dealing with combustion, they created the weakest flame ever seen in a laboratory environment, about one/two-hundredth that of a match, which would be significant since soot contributes to 60,000 premature deaths each year in the United States.

My question is this. We do have a future in medical research. Your own board of scientists came back to you and said that is a future for manned space research. The question is: If the Space Shuttle is grounded for a year or six months, what would be the impact on research, or do you foresee something even further down the road for the use of the Shuttle? And, secondly, if the space station is not serviced by the Shuttle regularly for a long period of time, what would the capability be to continue the use of those microgravity conditions? Or do any of our international partners have a vehicle capable of servicing, including assembly, the station?

Mr. O'KEEFE. Thank you, Senator.

If I could, just on the front, associate myself with your comments about the folks from East Texas. Unbelievable support. And folks like C.G. Macklin, who is the city manager of Lufkin, Texas, Captain Paul Davis, from the Department of Public Safety down there, unbelievable people who have stepped up in a way that is just truly heroic, and we are grateful to them. They have never been associated with the NASA family, and yet here they are contributing in a way that really is remarkable, and we are eternally grateful to them for their assistance as we've moved through this very difficult time in working through the challenges there.

The impact on station, to be sure, is a real difficult circumstance, given the fact that the next flight that was due in March would have been a crew rotation for Expedition 6 to be replaced by Expedition 7. To the extent that we are able to get answers to the current challenges that are underway that the investigative board is looking to. If we can get back to flight and resume flight operations, there should be no diminution of that support to station.

To the extent that that doesn't happen and the best scenario is not realized, we do have the capability on the Soyuz flights, which is the twice-a-year rotation of the emergency egress capsule. Three cosmonauts were due to be sent up—cosmonauts and astronauts—were due to go up in April. We're looking at what that crew configuration is, consulting with our International Space Station partners to determine the best way to configure the crew to use it potentially as a rotation capability for the folks that are aboard International Space Station now. Ken Bowersox, Don Pettit, and Nikolai Budarin potentially have the opportunity to come back aboard that particular Soyuz return vehicle and send a replacement crew up. We're looking at what those options may call for.

In terms of the long-term sustainment of the International Space Station for science, there's no question, between now and June there is an ample trove of science aboard the station right now that Ken Bowersox assures me, and, more important, Don Pettit, who is the science officer, says has got him occupied every single day and won't be a limiting factor between now and the time the summer rolls around.

Beyond that, there's no question, it would end up likely be a sustaining capability, because, in pursuit of the earlier commentary we just had, and conversation, there is a—the up-mass, or the capability to be able to lift the science experimentation in the middeck lockers and so forth that are aboard the Shuttle typically are what bring the scientific experimentation return or rotation for the International Space Station to that laboratory condition, and that would not be feasible to do. You can't get all that aboard a resupply vehicle like Progress, which is unmanned and for logistics and basic consumables. For everything you take out of it, it's that much less sustaining capability we have for the human beings aboard, and the humans are going to be the primary focus of our intentions. So, therefore, we would see a limitation and a diminution of the science focus that would be aboard.

But, for right now, it is positively stationed today as it was yesterday and will continue through the balance of this time as the most capable laboratory condition we have, and we are maximizing the science, and that sustains for several months to come. We'll have to make adjustments beyond that, if this goes beyond that period of time.

Chairman BOEHLERT. The Senator's time has expired.

Mr. Etheridge.

Mr. O'KEEFE. Thank you, Senator. I appreciate it very much.

Chairman BOEHLERT. Mr. Etheridge.

Representative ETHERIDGE. Thank you, Mr. Chairman.

And, Mr. O'Keefe, thank you for being here. And let me also associate myself with expressing condolences to the families and others, and also with the concerns that many of my colleagues have expressed today to make sure this study is independent. I think that needs to be done for the confidence of this Congress and for the American people for the future of the program, which I strongly support and think it's important to continue.

And let me say the people of North Carolina share with you greatly, because we will celebrate the 100th anniversary of flight this year in two of the four celebrations in this country. So we have a deep commitment to space and to flight.

I was in school on Monday, right after the Saturday terrible disaster. Children were concerned, obviously, as they always are. You mentioned earlier, someone did, and I think it's appropriate to cover it, because this is a great teaching tool, not disasters, but space flight.

In the 1960's, President Kennedy said we're going to put a man on the Moon before the end of this decade. We didn't know we could do it. We didn't know how to do it. But it spawned the growth of scientists and engineers that you talked about that were getting ready to age out. Don't you think it's about time we had another grand plan and decide we're going to put a man on Mars or some great planet? That may be above your pay grade, but someone needs to say it so we get another generation of excited young people to decide they want to get involved.

I know the scientists we have in NASA, which is a very small, elite, capable group are there because they were excited. But we need a bigger core.

Very quickly, because I have one more question I want to get to you. That is, beyond that—and I hope you'll speak to that—on March of this year or last year, the independent Aerospace Safety Advisory Panel reported to you that, simply stated, the panel believed that the repeated postponement of safety upgrades, restoring aging infrastructure, and the failure to look far enough ahead to anticipate and correct shortfalls and critical skills and logistical availability will inevitably increase the risk of operating the Space Shuttle. However, since then, I understand that NASA has cancelled planned upgrade projects, shifted funding for upgrading further out in time, and has indicated that it needed to do more studies of what the upgrades should be and how they would be undertaken.

Can you tell us if that's true, and, if so, why and how that will help improve safety?

Mr. O'KEEFE. Yes, sir. To your first question on big goals, you bet, the approach that the President, I think, has advanced as a part of our plan that as a part of the strategic plan and all the objectives therein is to develop those enabling technologies that would then permit the establishment of those big goals to be attainable.

And the two major limitations that I think we have got to beat down and be very, very thorough in our efforts to explore the technology opportunities to conquer is the ability get anywhere in a period of time and speed that would inform the research agenda and also assure that humans, when they go, can survive the experience for the full duration of that flight.

And as it stands right now, based on our current technology, just to get to the edges of this solar system would take us 15 years. That's an unacceptable period of time it would take. And assuming that any of the scientists, the principal investigators, the research focus, are still interested by the time someone would arrive there is one of our biggest problems, because things change an awful lot in the span of a decade and a half.

So our first objective, which you see dominantly in last year's budget and this year's budget, is how to beat and how to conquer the in-space propulsion power generation requirements we have.

The second dimension of that is to look very carefully at how we can assure that humans survive the experience. And, again, the degradation we see of the five expedition crews who have been aboard International Space Station for sustained periods of four to six months or longer is typically a physiological challenge, and we've got to figure out how to conquer that, because the amount of time it would take to roundtrip to anywhere that seems to be of curiosity that would be informed by research and scientific objectives, and we've got to be sure that the folks can survive that experience.

So that and the radiation effects, all those things, an intensive amount of effort that you see in the budget proposal before you, is concentrated on trying to conquer those kinds of limitations and understand what it would take to assure a safe roundtrip activity in that regard.

As it pertains to the second point you raised of upgrades, what we have proposed in the budget amendment that came forward last November 13th of 2002 was a direct consequence of recommendations from both the General Accounting Office, the Safety Advisory Panel, all the different external groups that we have had reviewing what we do, have suggested that the longer-term Shuttle requirements, to the extent we want to sustain that capability, require that we look at modernizing and upgrading those capabilities each and every time, as we do in the Orbiter Major Modification Program. What's in the proposal for fiscal year '03 that the Congress is still deliberating on now and for the '04 program that the President just submitted a week ago is a very specific plan that would provide for those increases necessary to sustain this capability through the next decade, primarily for lift capacity of those requirements, as well as crew transfer capabilities to and from International Space Station and elsewhere. That's as maneuverable as we can make it.

Chairman BOEHLERT. The gentleman's time is expired.
Mr. O'KEEFE. I appreciate it very much.
Chairman BOEHLERT. Mr. Smith.

RE-EVALUATING NASA'S MISSION

Representative SMITH. Thank you, Mr. Chairman.

I don't have any question beyond those that have already been asked, but I do have a statement I'd like to make and then I'd welcome Mr. O'Keefe's comments when I finish.

Mr. Chairman, I have long supported our efforts to learn more about the universe around us. In fact, I've always thought that a great rallying cry would be "one percent for space." That is, we should commit one percent of our national budget, or about double what we now spend, on scientific discoveries beyond the bounds of Earth.

The *Columbia* disaster, though, has made me question not our financial commitment, but the nature of our space initiatives. Perhaps we should re-evaluate some of our missions.

Launching astronauts into an inherently dangerous environment is always risky. Such efforts should be made only when the results justify the sacrifices. That may mean NASA undertakes fewer manned missions and more unmanned ones.

From what I read and hear, astronauts on the space station spend most of their time on maintenance and conducting experiments that could be performed by mechanical means. Of course, human judgment sometimes is indispensable, so there always will be a need for manned missions. But robotics should be employed more often. They can achieve our scientific goals more cheaply and with less risk to astronauts' lives. In other words, can we justify decades of repetitious yet demonstrably lethal roundtrip Shuttle flights to a space station that has not met expectations?

It's tempting to aim farther, at an inhabited outpost on Mars, for example. However, that endeavor could be one hundred times or maybe one thousand times more dangerous than a Shuttle flight. And through the video eyes of a Martian rover, we've seen what the planet already looks like up close.

Nor are prospects great for exploring our cosmic neighborhood, considering the distances involved. Our fastest spacecraft can travel a hundred times faster than a bullet. Yet even at that incredible speed, such vehicles would take 100,000 years to reach the nearest stars.

There is a way to reach across the expansive space, seize the public's imagination, and learn more about ourselves and the cosmos, and that is to search for signs of life elsewhere in the universe. A new generation of telescopes launched into space would be able to identify life on worlds orbiting nearby stars. Other types of telescopes could detect radio or light signals from distant civilizations.

Discovering the probable existence of life "out there" would cause more excitement than any news event in the history of humankind. It would certainly be a showstopper. And the possibility of extraterrestrial life, a show-starter for our next major space program. Searching for something more than microbes, for planets like the

Earth, and for other sentient life forms could lift our faces again to the heavens with hope and expectation.

Shuttle flights using new space planes would service these telescopes. No doubt, such operations will often require the good judgment and capable skills of astronauts. And astronauts who have gone before, such as those aboard the *Columbia*, will have laid a foundation from which we can push off seeking to discern the secrets of the universe.

Mr. O'Keefe, I'd like for you to respond to the idea, if not my specifics, then the general idea of reframing the mission that we have in mind. And if you have other suggestions, I'd be happy to hear them. I mentioned one for seizing that public's imagination and yet I'm learning more about the universe, and I'd be happy for you to respond to those suggestions.

Mr. O'KEEFE. Yes, sir. Well, no, I thank you for the very thoughtful commentary. And those are, I think, the same kind of issues we're wrestling with, in terms of what our appropriate strategy should be.

It's not an either/or proposition; it's a capability in which you build on the robotic unmanned, non-human intervention of capabilities that you can deploy and then utilize human requirements when necessary.

Again, the best example I've come across in my short one year of tenure at NASA is the Hubble Space Telescope. It is just a remarkable instrument today. It's something that is rewriting the astronomy books. Folks are just marveling at the capacity and the imagery that's coming back from the information from the Hubble Telescope is not only the new imagery that we're getting, but it's also informing the archival data that was collected in the last few years that suddenly now makes more sense because of the information we're getting today that now puts that in a different context and makes it more spectacular.

The reality is, that fantastic instrument would never have worked had we not had the capacity to launch a Space Shuttle and send folks to the Hubble Space Telescope to make the adjustments to correct the problems, which was, again, roundly considered to be space trash 10 years ago. This same instrument that was roundly, you know, dismissed as a mistake, has turned into a marvelous piece of machinery.

Representative SMITH. I just think we need more instruments just like that, but a bigger program and a more expensive——

I know my time is up. Maybe we can discuss this further later one.

Mr. O'KEEFE. Yes, sir. But I hasten to add, as well, that what you see before you in the current configuration of International Space Station is the same kind of example. This is a work in progress. We are six flights away from still achieving what is a core configuration. It had been planned to be resumed by this time next year, is where we'd be in that configuration, to build the scientific laboratories around it. But, at present, we're still looking at an amazing laboratory condition that is, in contrast to your characterization, sir, with all due respect, they are spending a lot more time on the science on these—and, matter of fact, the last two expeditions, Peggy Whitson just returned as the first science officer aboard, as we have transitioned from this engineering phase to one that's more intensively focused on the science.

It does take a lot, at least two folks to maintain it. No question about it. But it is—as we are able to build the crew capacity and focus on the scientific objectives, once we have reached a configuration that would permit that full use of the laboratory, it's going to be, I think, the same result that we saw out of Hubble in the long-term, which is going to yield the kinds of breakthroughs that we never dared imagine.

As humans, we are impatient. We want to see it now. And yet, at the same time, I think the persistence that we're trying to exert is to say, and the perseverance, is to make sure that we have that capability so that those kinds of revelations, like what we see today coming from our Hubble in our stick-to-it-iveness over this past decade yields the return we'd hope for. And it is today, and it will tomorrow, if we keep this up.

Chairman BOEHLERT. Thank you very much, Mr. O'Keefe.

Mr. O'KEEFE. Thank you, sir. I appreciate it.

Chairman BOEHLERT. You will note, and you're a frequent witness our panel, the House members are particularly skilled at time management, because we operate under different rules. And so they've developed the knack for asking—using all their time to ask their question, and then obviously we'll give you the opportunity to respond, because no question should go unresponded to. But we're going to stick, as much as possible, to the time limitations in the interest of all concerns.

The Chair now recognizes another skilled practitioner of the art of questioning, Ms. Jackson Lee.

Representative JACKSON LEE. Mr. Chairman, what an introduction.

(Laughter.)

Representative JACKSON LEE. Let me thank the Chairman and the Chairman of the Senate Committee for this opportunity.

And, to Mr. O'Keefe, you have shown the proudness and wisdom that we have seen over the last two weeks, compounded, of course, or matched, with your compassion and love for the NASA family. My sympathy to the extended family, and specifically to the families of the astronaut, *Columbia* seven. A local newspaper called them "Astronauts, The Heroes Next Door." And I do want to announce to you, and we're very pleased, that now almost 80 members of Congress in H.R. 525 have joined us to give them the Congressional Gold Medal, which is the highest civilian honor. And we look forward to NASA supporting us. We believe that we can move this legislation quickly. We are going to call on our Senate colleagues in that, and we are very grateful to Republicans and Democrats who have signed onto this legislation very quickly.

I think the important message that I'd like to convey in my brief time is that the Shuttle will fly again, and that the challenge should be, for lack of a better term, that it flies sooner rather than later, sooner than two-and-a-half years, sooner than three years.

You've heard this before, so let me focus on it again. I think it is extremely important that we have the Admiral's committee expanded, and I think it you need to consider the likes of a Nobel laureate, an academic, some industry engineers should be considered, some advocates of aviation. And, I believe, after 9/11, depending on their desires, family members or representatives should be considered to be part of this committee, because our job is to instill confidence in the employees, in the families, in the astronauts, not necessarily in that order, and the American people and this Congress.

And I'm reminded of the Rogers Commission that had a subsection, the Silent Safety Program. And I assume if I was to read that, it would again comment on the issue of safety.

Let me bring these points to you and tell you what I'd like to hear, whether it be in writing or you'll be able to say it now. I want actual dots, a road map, to lead me from —or to the conclusion, whether it be a conclusion that is not popular, that budgeting did not interfere with the safety of this program. I want an actual—we don't want to be presumptive, we don't want to speculate, and we don't want to be afraid of saying "mistakes."

I want to be able to understand about the frozen foam that fell under the underbelly and then hit the left leaning wing, and how we can speculate that that didn't count, when we had a report in 1994 from Stanford and Carnegie that suggested 15 percent of the tiles could count for 85 percent of the damage. And I understand an engineer in 1997, most recently, said that he thought debris falling might have an impact. And as we all know, this goes on its belly, and, therefore, it's possible for debris to hit while it was enroute or while it was in space. So I'd like the direct lines to that.

I'd like to also say that an orbiter Shuttle is excellent, but I'm very interested in payload. I think it is a valuable part of what we do, and I would not want to just have a vehicle that transported human beings, because I want research to be able to be done, because we're saving lives. And if you can comment on that idea, because I understand that we're pushing forward with the research on the orbiter, I am certainly excited about that, but I want to make sure we can carry a good payload so that that research, that vital research, can be done.

If you can comment on the fact of the icicle that fell, I call it that, and why we could suggest, or should suggest, that that was not a problem, and that you will instruct, or however the instructions are, to this commission, the committee headed by Admiral Gehman, that he will leave no stone unturned and that we'll be able to track or follow his tracks.

Budget cuts that did occur did not have an impact. An aging vehicle did or did not have an impact, 30 some years old.

[The information follows:]

The Gehman Board has been chartered to look at all elements of the accident to determine the cause and to recommend specific actions to be taken in order to fix the cause and move forward. The members of the Board have a significant track record of experience in serving on aviation accident investigation boards. They bring knowledge, expertise and independence to the table. We anticipate that the Board's findings will outline a roadmap for all to see.

It is too early for NASA to comment on potential causes of the accident. The Gehman Board is addressing all possible scenarios of what could have caused the accident and as such no scenarios have been ruled out.

NASA commissioned the 1994 Stanford/Carnegie report cited, *Study of Risks to the Shuttle Tiles*. The study concluded that ice on the Shuttle's external tank was a major source of debris that could damage the tiles. The study also concluded that the most vulnerable areas of the Shuttle were the undersides of the wings and the crew compartment.

NASA has taken a series of steps to reduce risk associated with this type of damage. We have: improved the process for inspecting, maintaining, and repairing Shuttle tiles; improved launch commit criteria to reduce the chance that ice falling off the External Tank could damage the Shuttle's thermal protection system during launch; and changed the formulation of the thermal protection system for the Solid Rocket Boosters.

There is space debris that could potentially pose a threat to the Space Shuttle. At this time, we don't know whether Columbia was struck by space debris. One of the teams supporting the Gehman Board is assessing the potential hazards that Columbia might have contacted during its mission.

The Shuttle serves multiple purposes. These purposes include being a research facility, transportation for payloads to be deployed, assembling/servicing/supplying the International Space Station and crew rotation for the ISS. In recent years, the primary mission for the Shuttle has been the assembly of the ISS, specifically, bringing hardware elements on-orbit for assembly as well as research payloads to conduct research on board the ISS. The Shuttle combines the payload carrier capability with the human interface necessary for successfully assembling the most sophisticated engineering marvel ever attempted.

It is too early for NASA to comment on potential causes of the accident. The Gehman Board is addressing all possible scenarios of what could have caused the accident and as such no scenarios have been ruled out. The Board has convened press conferences to convey information to the American people relative to their investigation. The Board's charter and membership have been posted on the web for ease of access by any individual. Admiral Gehman has been rigorous in assuring the independence of the Board.

Chairman BOEHLERT. The gentlelady's time has expired, proving the point.

Representative JACKSON LEE. And I thank the distinguished Chairman.

Chairman BOEHLERT. Mr. O'Keefe, you have.

Representative JACKSON LEE. I had come to an end of my sentence. If the——

Chairman BOEHLERT. Well, thank you very much.

Representative JACKSON LEE. If he——

Chairman BOEHLERT. Mr. Administrator——

Representative JACKSON LEE.—could comment briefly, I would appreciate it.

Chairman BOEHLERT.—you'll have the opportunity to respond, by all means. We would not——

Representative JACKSON LEE. Thank you very much, Mr. Chairman.

Chairman BOEHLERT.—cut off——

Mr. O'KEEFE. Thank you. Very probing questions, no question, all of which I think we should find the answers to, we must find the answers to, to understand exactly what happened to the *Columbia*. And the investigative board, when they come to conclusion on each of those points, we positively will be guided by it.

But let me offer the following observations. Everything we've seen on the budget and the resource profile for Shuttle would suggest that concurrent with improvements in efficiencies, there were also improvements in performance. And to the extent that doesn't bear out, to the extent there are systemic problems that the investigative board concludes led to this tragedy, that's the kind of thing we'll have to fix, too. And their charter is broad enough to cover that range of issues. We're not just looking for a technical finding of what happened on this one flight. Anything else they want to observe, they are free to do so, and we'll be guided by their view.

The operational problems, again, I really want to avoid any favorite theory of what it is that could have contributed to this. We have closed off no branch in this fault-tree analysis, if you will. We want to make sure that everything is analyzed, every possible thing that could have contributed to this, to include the foam pulling off the External Tank, whatever. All of those are theories that, again, are current. They're certainly plausible, and we're going to be guided by the investigative board's conclusions of what their ultimate contribution was in these kind of cases. So I really want to be sure we're not shutting off any of those avenues, but, at the same time, not pursuing one we think is more likely or favorable or not.

In my limited experience with dealing with crisis circumstances or management of situations where you're responding to incidents, typically one of the variables that occurs, not always, but many, many times, the initial evidence proves to be not nearly as illuminating as it was when it first came out. And so rather than tracing or chasing what turns out to be a blind alley, and, therefore, foreclosing and letting the trail go cold on all kinds of other options, we're trying to maintain an even-keel approach of being sure that we not go out and favor one favored approach versus another, and to let all the evidence, let the facts speak for what ultimately occurred in this case. So we're avoiding that.

And I, too, am looking forward to the assessment of the investigative board's conclusions about how that particular item—whether it was the contributor or not. And that's what I'll be guided by as we move along.

Lastly, your observation that you asked the science content. Absolutely, we have to really maximize that. That's the—the risk that we deal with each day, those seven heroic folks who went aboard the *Columbia* were making a contribution to, ultimately, the science and research objectives. We have to have a equally intensive and disciplined approach about what we would ask them to risk their lives for, and be equally serious about it for every one of those cases in which we ask folks to venture off to do these things.

And so we've got to be as disciplined as they are in their training of assuring that the science and the research yield we think could come from this meets that same test and standard. And that's what we're about, that's what I think we're attempting to do on International Space Station. We're trying to build that capacity to yield those kinds of breakthroughs that would never be possible were it not for that facility that can't be duplicated here on Earth. So we continue in that pursuit. We are completely in agreement on that, really, imperative to be that serious about it. And I thank you for your observations.

Chairman BOEHLERT. Thank you very much, Mr. Administrator.

Representative JACKSON LEE. Thank you, Mr. Chairman.

Chairman BOEHLERT. Mr. Feeney.

Representative FEENEY. Thank you, Mr. Chairman, and for Chairman McCain and the Senate for their hospitality today.

Mr. O'Keefe, it's great to talk to you. I know it's been a very difficult weeks for you and your staff, and I would wonder if you could comment on a couple of thoughts that I had.

Number one, within about 15 minutes of the *Columbia*'s failure to land on time at Kennedy Space Center, there was a contingency plan put into place, and I'd like to know what, if any, steps you took upon assuming the control of the administration at NASA with respect to reviewing, familiarizing yourself with the contingency plan and what you found.

And, secondly, I was struck by the portion of your testimony when you spoke. I think you used the phrase "the ethos of safety at NASA," and I think you included the contractors who work for NASA, as well.

Not long ago, you and I sat as we hoped for a liftoff. We didn't get one that day. But in the audience, a young man was introduced by the name of David Strait. I wasn't familiar with his name. I thought, by the reception he got from the people there that day, he must be a rock star or a TV star. I don't watch much TV or listen to much music, either.

But I wonder if you could comment on the fact that the people involved in this program are passionate about it, what your view of their professionalism is, and what we can do as we go through this very difficult process of the investigation, the fix of the problem, and hopefully creating a new vision of space to enhance that esprit de corps during some tough times.

Mr. O'KEEFE. Thank you, Congressman.

The issue of the contingency plan, no question, this was something that occupied my interest from the first hour that I was privileged to be in the capacity as administrator at NASA. On the first day I walked into the office, I asked folks to please take off the shelf whatever it is, whatever that plan is, of how we would respond to a disaster like *Challenger*. And I asked that that meeting occur within an hour of when I requested it so that nobody would have to feel like they have to run off and make something up, that they'd just pull off the shelf what was there. And we sat down and reviewed that plan, and it clearly is of the origin from the *Challenger* circumstance, no question, an awful lot of the contingency planning efforts that went into it. And I reviewed it in great detail, to assure myself that I would have some working familiarity with it.

Then I asked our senior leadership folks to then benchmark it against the only other community I know of that is as equally obsessed with safety as NASA is, which is the nuclear reactors community, the naval reactors community, the legacy of Hyman Rickover and all the folks who have, over 40-plus years, have operated safely over 125 billion miles, they say, of safe operations of nuclear reactors.

So we imposed upon Admiral Skip Bowman, who is the head of naval reactors today, who is Rickover's successor several times removed now, to benchmark, help us benchmark, relative to the approach of what they use as their contingency planning efforts. Over the course of the next six months, we were able to compare notes, if you will, upgrade our plan, he upgraded his based on the way we do business, too. We both benefitted by the exercise. The contingency plan was updated and republished in September of 2002. We ran a simulation of it in November, secure in the confidence we'd never have to use it. But, nonetheless, it was organized that way.

And specifically, all the folks who were to be on the Independent Investigation Board, named, identified by position, were notified so that everybody understood the procedure. All the folks internally were exercised on the activity. And we were confident we'd never have to really worry about using something like this.

On the day it occurred, at 9:29, Bill Readdy had it with him, as did all the senior officials at Kennedy, opened up the first page, and we went to item one on that contingency plan. And it was the saddest moment I can ever recall, to be followed by the most tragic moment I've ever experienced in my life, which was to face the families of these crew members, but to tell them and to reassure them that we are working through this as diligently as we know how to find what caused this, what are the answers, how are we going to fix this, and assure that we pursue the same dream that their spouses, loved ones, fathers and mothers wanted to see pursued.

And so this plan is as good as we know how to put it together. It is that legacy. It has been really worked as smart as we know how.

And the safety ethos, if I can get to slide 23 real quick, to be sure, the example you cited, the fellow you talked about, David Strait, notice on the lefthand of this chart, there's—you can barely even see it—there's a hairline facture of no more than about an inch and a half that was enough to ground the entire orbiter fleet for four months until we knew what was the source of that problem. How did it happen? It isn't supposed to have shown up in anything. We stopped all flight operations, made the repair area to it that you see on the right over the course of that time, after they had run many, many simulations of this to figure out what the right answer was. And yet there's a bead weld right above it that was there without any consequence since the day this orbiter was first put together. And this effect was made on all four of those Shuttle flights.

So even something as minuscule as that that David Strait, with 20–20 vision, noticed as he just went through his routine business, like they all do, of inspecting the orbiter, every square inch of it after every flight, noticed that seemingly innocuous problem, and that was enough to ground that fleet for four months.

Chairman BOEHLERT. The gentleman's time has expired.

Ms. Lofgren.

Representative LOFGREN. Thank you, Mr. Chairman. I'd like to thank you, as well as Senator McCain, for convening this hearing. Over the next few months, I think all of us will be asking tough questions related to the loss of the *Columbia*, as well as the future of the space program.

But it's appropriate that first our country has paused to reflect on the heroism of the seven astronauts who gave their lives so that the dreams of humans reaching for the stars could live forever. And my thoughts and prayers are with the families of those we lost, as well as to the extended NASA family. And I know from my own experience that part of our NASA family at home, at NASA–Ames, is very much grieving with the rest of the country over this loss.

You know, I also believe that this committee is going to focus on asking difficult questions that relate to how we're best able to resume our quest to explore space, and that's really the best way to honor those who were lost, to ask those tough questions and to find answers.

Clearly, we're not going to find the answer to the specific technical issues relative to the *Columbia*. We've got technical people to do that. But it is appropriate for us to examine our own actions and to question each other about the policies and whether those policies had any impact on the risk that was inherent in this flight.

Clearly, at least into the foreseeable future, space flight will be risky. And we know that the percentage of odds right now is—although initially we thought that the use of these vehicles would have a risk of one in one-hundred-thousand, it's down to now one in fifty-seven if you just look at the records. And so we need to—I'm a believer in human space travel, but we need to make sure that we're doing our part to minimize the risks.

Now, I was late for this hearing, and I wish to apologize. It's been a big science day. We spent all morning on stem cell research in the Judiciary Committee, and I have hopes that science will do better in the Science Committee than science did in the Judiciary Committee today.

But one of the questions I have for you, Administrator, is; as we look at what we should do to make sure that the risks are minimized, were there any safety upgrade proposals ever made to you, either as Administrator or in your prior life over at the OMB, that you did not support? And if so, what were those recommendations, and why did you reach the conclusion that you did?

Mr. O'KEEFE. Not that I'm aware of, but I certainly will review the history of both of my capacities in the course of this Administration and ascertain the dates of when there were any deferrals or anything else of any upgrades that would be categorized as exclusively focused on safety. So, to my knowledge, we have not done so.

The only issues I'm aware of is an Electric Auxiliary Power Unit upgrade that had been planned that was determined to be technically deficient and wasn't—you know, so, in other words, no amount of money we threw at it was going to yield its performance in the manner in which it would contribute to not only efficiency but also safety characterization—that was deferred. And we're now re-examining to figure out how we can pick that up or continue it in the future that would yield the performance requirements we know of.

But we will go back, and I will submit for the record any other changes that were made during the course of— well, since Inauguration Day 2001, and if there are any changes that have occurred in that time, we'll certainly report those.

[The information follows:]

Administrator O'Keefe has not rejected any Shuttle upgrade proposal as NASA Administrator or during his tenure at the Office of Management and Budget. Since Mr. O'Keefe has been the NASA Administrator, the Administration prepared and submitted to the Congress in November 2003 an amendment to the FY 2003 budget request to increase the funding for upgrading the Space Shuttle system by approximately $660 million for the FY 2004-2008 timeframe. The budget amendment recognized that the Space Shuttle would be the workhorse for International Space Station transport through at least the middle of the next decade.

In 2001, the electric auxiliary power unit (EAPU) was experiencing technical difficulties, cost growth, and schedule delays. This led NASA, with the endorsement of the NASA Space Flight Advisory Committee (SFAC) and the NASA Advisory Council (NAC), to cancel the project in mid-2001. In the FY 2002 Operating Plan, the Space Shuttle program cancelled or deferred several upgrades because of cost growth or technical immaturity. In the Operating Plan, reviewed by Congress, the funding made available as a result of these actions was then applied to Space Shuttle operations to accommodate operations cost growth. These actions did not affect safety. In September 2002, NASA's Office of Space Flight canceled the supportability upgrade for the Checkout and Launch Control System (CLCS). The decision was based on: unforeseen development difficulties with software, uncertain confidence in meeting schedule, and significant growth in development and projected operations costs, as well as the fact that this upgrade would not have been significantly more capable than the existing Launch Processing System.

Per the latest update to NASA's Integrated Space Transportation Plan, which extends the Space Shuttle's operational life to the middle of next decade, the Administration's FY 2003 budget amendment increased out-year funding for the Space Shuttle program. This increase provides for an additional flight in support of the ISS and an increase in funding for upgrading the Space Shuttle system of approximately $660 million for the FY 2004-2008 timeframe, through a Shuttle Service Life Extension Program (SLEP).

Representative LOFGREN. So you'll go back and review the record and take a look at your—obviously, hindsight's 20–20; we're all doing that in terms of our own activities —what you recommended both at OMB and in your role as NASA Administrator. And I know my time——

Mr. O'KEEFE. I'll do my best.

Representative LOFGREN. Thank you very much. And my time is up, Mr. Chairman. I appreciate it very much.

Chairman BOEHLERT. Thank you very much.

The Chair recognizes the Senator from Washington.

LESSONS FROM THE *Challenger* INVESTIGATION

Senator CANTWELL. Thank you, Mr. Chairman. And thank you, Mr. O'Keefe, for your diligence today. I've been to several meetings since I first checked in here this morning, and I think you've had a total of a five-minute break. So thank you for your diligence in answering these questions.

I don't think any of my colleagues have asked specifically about the lessons learned from the *Challenger* inquiry. And I don't know if, in this current configuration of the *Columbia* inquiry, you think that we have a sufficient independent scientist on this review team.

Mr. O'KEEFE. Yes, Senator, very specifically, the contingency review plan and the activation of an investigative board is a direct outgrowth and a direct education from the *Challenger* accident. So what we put in motion on the day of the accident was something that was a lesson learned from *Challenger*. It was part of the Rogers Commission recommendations of how we would proceed in various cases, and this is an outgrowth of one of their concerns, which is how to get ahead of these cases as quickly as possible.

The investigative board was appointed the same day as the accident. So, as a result, that was a very clear result of the lesson learned that came from that.

So an awful lot of what we attempted to do here is to build on that experience and assure that we have a result, in this case, that is driven by our interest in absolute dedication to finding the answers to what caused the terrible tragedy, find the solutions to it, the fixes, and get about the business of getting back to safe flight to support the folks aboard International Space Station.

Senator CANTWELL. So who is that independent scientist, then, on——

Mr. O'KEEFE. Oh, I'm sorry. The independent—there is—I spoke as recently as last night to Admiral Hal Gehman, who is the chairman of the current board, who has five different folks he has in mind, I'm told, who are physicists, scientists, etcetera, that he is considering to propose for addition on the board. I have advised him whoever he wants to put on that board that will expand his expertise, that will improve the independence of the board, help its objectivity, we will do it without qualification and without hesitation.

Senator CANTWELL. Well, I am struck, reading last night and this morning, the Feynman minority report to that report that I think finally got in as an appendix in which Mr. Feynman was very critical of discrepancies between engineers and managers as the probability of failure. So you had engineers having studies and analysis saying that maybe the risk is a lot higher than what the high-level managers thought. The criteria used for flight-readiness reviews often developed a gradual decreasing strictness, "If the Shuttle had flown with it before, chances are it worked before, so let's just—let's not think about the variations that might happen."

The Feynman Report even said NASA might have exaggerated the reliability of product, that there were variations in models, that NASA was more of a top-down system in testing the entire system instead of testing the individual properties and limitations of the material within the Shuttle to the degree that I think the report was quite critical of NASA, in the sense of maybe even coming to Congress and catering to us in the public relations expectations that were there by the public, instead of relying back on this basic engineering information.

So I guess my concern is, is that the panel, as I see it now, doesn't have that Feynman voice, and I think what we really do want to do here is make sure that we are not engaged in that PR battle, but understanding how we really do build the NASA systems of the future——

Mr. O'KEEFE. Right.

Senator CANTWELL.—with more reliability and predictability based on those materials.

And so I'd be very interested in how those recommendations were actually implemented, because I have a feeling we might find the same circumstances are true here. We're going to find out that there is some material property limitations that were discovered in some report written by some engineer that somewhere along the way got translated into "not as big a risk" and, thereby, the disaster that we've all been dealing with. So I appreciate your attention.

Mr. O'KEEFE. Well, thank you, Senator.

That may be. And if the investigative board comes back and points to the systemic problems that we have, we positively will take that as a firm recommendation and go fix it.

What I can advise, though, in my one year experience, with no prior experience with NASA at all—this is my first year at it—and reading the *Challenger,* the Rogers Commission Report, I was a Senate Appropriations Committee staff member on the day *Challenger* blew up, read the report thereafter, and that's my only association with that up until a year ago, is—what I see is a different agency than what existed 17 years ago, in that sense.

One of the observations they made, Feynman's view, I think, was exactly right, he pointed to the difficulty of that chain of command and how it gets altered. This whole process I've witnessed, and I'm advised I'm the first Administrator to have attended what is called a Flight Readiness Review two weeks before a launch, it resembles a room like this, with everyone that you could possibly imagine associated with this activity, all of which are empowered to raise their hand during the course of a full day, sometimes two-day, review of every single technical issue. And if they disagree with the way it was presented, they disagree with the conclusion, they immediately raised their hand, and the issue was then put to the side to go work the conclusion of it. That didn't exist prior to *Challenger.* It was all done by telephonic tag-up occasionally. This is a in-the-room, everybody there associated with the activity.

There was a big difference at that time, where schedule drove everything. They were looking to get to a flight rate of 20, 30 flights a year. We're operating on a five- to six-flight-a-year approach.

And as we discussed with Congressman Feeney a minute ago, and I appreciate your bringing this out, we stopped flight operations for four months over a hairline fracture found on not the orbiter that was scheduled to go up, but one that's in an Orbiter Processing Facility. Everything ended. The engineer, the inspector, that noticed that, that stopped operations cold for four solid months.

The disjointed chain of command, that doesn't—I don't see it. We have astronauts, former astronauts, who are in capacities as high as the Deputy Administrator, the Associate Administrator for Space Flight, the Associate Administrator for Safety and Mission Assurance, all three are former astronauts. That didn't exist prior to *Challenger.* None of those positions included anybody with that kind of background.

And all the way through this process, every single one of those managers are empowered, are expected, there's a responsibility that each of them feel they have, to stand up and be counted and stop all the operations until there is any issue that has been left unresolved, beaten to ground truth.

So what I've seen—and, again, from an objective opinion, I think, until a year ago, and now I'm steeped in it, there's no doubt about it—but until that time, unfamiliar with it other than what I read in the *Challenger* Rogers Commission Report, as well—would tell me this is a different place than it was then.

All that said, we will be guided by what the Gehman Board comes back and tells us was the problem here. And if it was systemic problem, we're going to fix it. If it was a technical problem, we're going to fix that, too. There is nothing I can imagine that's not on the table, and I have no bias against any finding they could possibly come up with that wouldn't otherwise contribute to the solution in this particular case. We are going to act on that without reservation.

Chairman BOEHLERT. Thank you very much.

Senator CANTWELL. Thank you, Mr. Chairman.

Chairman BOEHLERT. Mr. Moore.

Representative MOORE. Thank you, Mr. Chairman. Thank you, Mr. O'Keefe, for staying.

Yesterday, in the *Washington Post,* it was reported, I believe, that you said that—you defended the way you set up the accident investigation board, arguing that you proceeded the way you did so that it could launch an investigation immediately with members who were already well briefed on Shuttle operations. I guess my question is, Mr. O'Keefe, how important it is that we launch an investigation immediately, or should we take a more slow and diligent approach, as I think you said to Ms. Sheila Jackson Lee recently? I'm just—I'm asking what kind of investigation do we need here? What's, in your opinion, the best approach?

Mr. O'KEEFE. Yes, sir, thank you.

I believe what I tried to say—and if I was inarticulate, I apologize—was to say we developed a contingency plan to activate an investigative board so that they could act immediately, get on with the task immediately. They are not—and if I said this, I am in error, and I need to correct the record—they are not conversant in Shuttle operations.

There is only one member of the group who is even vaguely familiar with NASA operations. The rest of them have had no experience with NASA at all. The chairman of board is Hal Gehman, a United States Navy admiral, retired, who never had any association with NASA. I don't even think he ever attended a launch before. All of the other members of the board are from the FAA, the Department of Transportation, the United States Navy, the United States Air Force, all folks dealing with safety, mission assurance, flight certification, etcetera. I don't know if any of them have even toured a NASA facility. They spent the past week in Shreveport, Louisiana, and have now, just now, this past Friday, arrived at Johnson Space Center, and went through the simulation of what a re-entry is like, astronaut simulations they do. None of them have ever been through that before.

So Admiral Gehman has advised me that part of what he's done this past week is get up to speed on what he calls "Shuttle 101," just to understand what the lingo means and what the acronyms are. They are, nonetheless, were available up and running and talking to each other as early as seven-and-a-half hours after the accident. On Saturday afternoon, at 5 p.m., they were already identified and ready to go.

So at least we picked up the time that gave them the opportunity to then become more familiar with the processes, the evidence, the facts, the data, so that they could get about the business of investigating as quickly as possible before the trails go cold.

That's the approach that—you know, certainly, there's no one-size-fits-all approach to this, but it certainly was one of the most effective ways to get moving.

Slide 33, if you would, at least this is what happened two days after the event. That's the folks that got there, and they're talking to FEMA, you know, managers on site, you know, it at least gave us an opportunity to get moving, as opposed to sitting around thinking about who should we pick, when should they go.

Representative MOORE. In terms of a realistic time frame, then, what might Congress and the American people expect us to—when you see this investigation really getting underway, and I know you can't predict what results we're going to find. I assume you can't predict what results we're going to find. But what time frame are we looking at, Mr. O'Keefe?

Mr. O'KEEFE. Well, the guys that's right on the other side of the fellow with the FEMA jacket on, right ahead of him, is Admiral Hal Gehman, and he can answer that question better than I can.

Representative MOORE. All right.

Mr. O'KEEFE. I wouldn't even presume to suggest when he's going to finish. And he has—there's no amount of time that's necessary that we think is appropriate to go out and find out what happened here. We're going to be guided by he and his board's view of exactly what occurred, and there is no time limit on that.

Representative MOORE. Very well, thank you.

Mr. O'KEEFE. Thank you. I appreciate it very much.

Chairman BOEHLERT. Thank you very much, Mr. Moore.

Mr. Administrator, I want to thank you very much. Before we wrap up, I just want to touch on a couple of more points. I want to bring clarity to a very important issue.

I think it should be self evident that the Congress is committed to the proposition, on a bipartisan, bicameral basis, that we need to strengthen the evidence supporting the assertion that the *Columbia* Accident Investigation Board is truly independent. Now, I know the message has been sent, and I think it's been received, and I know it's been heard, and I want some assurance that it will be heeded. So I'd like you to visit that a little bit more and comment.

I carefully listened to you as you said you'll consult with Admiral Gehman, but I hope you're hearing what Congress is saying. We are the ones, and others, too, insisting that we get some clarity to this issue. So can you address that a little bit more for me?

Mr. O'KEEFE. Yes, sir. No, I appreciate your patience on it, and I do not mean to equivocate in any way, shape, or form.

I share exactly the same objective, I think, as all Members here do, which is to determine what are the answers to this tragedy, what are the facts that led to it, and ultimately find out how we go about fixing it and getting back to safe flight.

And in that pursuit, I will not just consult with Admiral Gehman, I will advise. As soon as this hearing is concluded, I will give him a call back and say my clear understanding from the Members of this Joint Committee is that there are aspects of the charter that need further revision. Let's examine what those might be. And, to his satisfaction, we will make a change——

Chairman BOEHLERT. Well, let me give you an easy one right off the bat. I mean, number 10, provide a final written report to the NASA Administrator not later than 60 days. First of all, the 60-day time frame——

Mr. O'KEEFE. Sure.

Chairman BOEHLERT.—is totally unrealistic. But the report will come to the President, to the Congress, to the American people, and to the NASA administrator simultaneously.

Mr. O'KEEFE. Yes, sir.

Chairman BOEHLERT. All right.

Mr. O'KEEFE. I'll make that an alteration and suggest to him that that's exactly—and we'll go ahead and make that charter change, because, again, I've stated that.

There's a letter, too, that I—that's part of the record, as well—on the 60-day issue, that, when I commissioned the panel in the very first place, said, "Our contingency plan contemplated 60 days, but you take whatever time you think you need, Mr. Chairman, Admiral, to come to conclusion on this." I'll reiterate that. We will eviscerate the 60-day. It has no bearing. It was intended as part of the continency plan originally, but not envisioned to be used. So to the extent that there is any amount of time he needs, that's what he's got. I don't intend to impose anything different on him.

Chairman BOEHLERT. Well, that's the easiest one.

Mr. O'KEEFE. Sure.

Chairman BOEHLERT. We have, you know, some others that——

Mr. O'KEEFE. Well, actually, we'll revise whatever is necessary.

Chairman BOEHLERT. We want to deal with it clearly so that it's clear in our own minds that they're truly independent.

Mr. O'KEEFE. Yes, sir.

Chairman BOEHLERT. And if they decide they want to go down a certain path or they want to hire a certain expert, they don't have to march over to NASA headquarters to get approval.

Mr. O'KEEFE. Yes, sir.

Chairman BOEHLERT. The admiral and the *Columbia* Accident Investigation Board have the authority to proceed as they deem best——

Mr. O'KEEFE. Yes, sir.

Chairman BOEHLERT.—to get us the answers we are all demanding. And that's critically important.

Mr. O'KEEFE. Yes, sir, I concur. I appreciate it. Thank you.

Chairman BOEHLERT. Well, I think, as we come to the end, we've done several things today. First and foremost, the current status report. And obviously, this is a very dynamic situation, so events almost change hour by hour, let alone day by day. And so that was very important.

And we've started what I have characterized in my opening remarks as the national conversation, which we have to start, people talking to each other, not through each other, over each, around each other—directly.

We have affirmed the commitment to the concept of an independent board, and we've had agreement on the need for charter changes. I think that is very important.

Now, this is not the beginning of the end; it's the end of the beginning. And we always want instant analysis of immediate findings, and that's understandable. But experience tells us we learn the most from in-depth examination of more complete data.

So now we are in the fact-assembling phase, and all of us with responsibility in this very important assignment—the Congress, NASA, the *Columbia* Accident Investigation Board—are all going about the business of assembling the facts. Then all of us will have this database to look at and take care and caution as we go forward with our special responsibilities.

Obviously, NASA and the Accident Investigation Board will be focusing more on the technical aspects. Obviously, there's a need on the part of the Congress to focus more on policy as we chart the course for the future. That does not mean they are mutually exclusive. We'll be looking at each other.

I have been very pleased with the response I've had from Admiral Gehman in assuring us that Congress will be very much involved in all of the proceedings. I have been very pleased with the cooperation we've received from Administrator O'Keefe and his team. And I have been just impressed beyond any ability to adequately explain at the total commitment I find on the part of every single person involved in this procedure to get the facts. And let us be guided by the facts as we fulfill our important responsibilities.

Mr. Administrator, thank you.

The hearing is closed.

[Whereupon, at 1:30 p.m., the hearing was adjourned.]

Appendix 1:

ANSWERS TO POST-HEARING QUESTIONS

ANSWERS TO POST-HEARING QUESTIONS

Responses by the National Aeronautics and Space Administration (NASA)

Questions submitted by Chairman John McCain

Q1. In a Wall Street Journal *article dated February 11, 2003, a fundamental question of "What is the next step for manned space flight after the Shuttle?" was discussed. The article also talked about NASA's four alternatives for replacing the Shuttle program, three of which would be limited to servicing the International Space Station. The fourth option would develop a space craft that takes off like an airplane without the help of unmanned boosters. This could allow for servicing of the Station along with a number of other options such as a mission to Mars.*

Q1a. Do you believe that we are short-changing ourselves by proceeding with a vehicle that can only service the Space Station?

A1a. No. The Orbital Space Plane (OSP) is only one element of the Integrated Space Transportation Plan (ISTP), which provides the roadmap for NASA's future investments in space transportation. The principal benefits of the OSP include: (1) providing assured access to the International Space Station (ISS), (2) meeting the U.S. ISS crew rescue responsibilities, (3) improving the overall crew safety, and (4) providing a bridge for crewed space flight on future launch vehicles. The OSP Level 1 requirements are focused on meeting the first three benefits based upon the Agency's near-term needs. But by addressing the crewed segment, OSP would also provide a crucial building block for future human space flight vehicles under the ISTP. The ISTP will continue to evolve consistent with NASA's strategic objectives.

Q1b. While there were technical difficulties in the X–33 program which led to its eventual cancellation, were the difficulties insurmountable if the Nation had made a conscious decision to pursue a manned space craft for exploration beyond the Space Station and the moon? What were the technical barriers to the X–33 program?

A1b. The X–33 was intended to demonstrate, in flight, the technologies needed for a full-size, single-stage-to-orbit Reusable Launch Vehicle. As a technology demonstrator, it required dramatic breakthroughs in multiple technologies, including the development of composite liquid hydrogen (LH_2) tanks that were an integral structural part of the overall vehicle. Achieving single-stage-to-orbit, X–33 was also founded upon the principle that the commercial market would continue to grow. The X–33 program had reached the end of the time period specified in the cooperative agreement between NASA and Lockheed Martin. When faced with the decision of providing additional funding to continue the X–33 vehicle after the failure of the LH_2 tanks, the Agency decided the X–33 would have to compete for funding with all other reusable space transportation system development efforts under the SLI program. This decision was communicated to Lockheed Martin prior to the completion of the original cooperative agreement, which ended on March 31, 2001. The X–33 proposal was not selected for award under the SLI NRA 8–30 Cycle I competition (May 2001).

There are multiple vehicle configurations that would be considered if the Nation made a conscious decision to pursue a manned spacecraft for exploration beyond the Space Station. Unless dramatic progress is made in the technologies necessary to achieve single-stage-to-orbit, it is likely that the final configuration would be different than the X–33 design.

Q1c. The Wall Street Journal *article also implied that if we shifted to capsules launched on expendable launch vehicles, it would be an admission that the Shuttle program never really made much sense. Do you agree with that implication?*

A1c. No. The unique capabilities of the Space Shuttle have enabled a broad range of missions that would not have been accomplished without it, including the Hubble Space Telescope deployment as well as repair and maintenance missions, satellite servicing, and construction of the ISS. The Space Shuttle remains the Nation's primary means of transporting crew and cargo to the ISS.

Q1d. Many are calling your space plane concept a step backwards because it is half disposable and not fully reusable. What's your response to these claims?

A1d. OSP does not replace the Reusable Launch Vehicle (RLV). The RLV architectures always consisted of NASA Unique Systems (renamed OSP) and booster systems. The OSP represents a step forward for the Agency, offering the following benefits: (1) providing assured access to the ISS, (2) meeting the U.S. ISS crew rescue responsibilities, (3) improving the overall crew safety, and (4) providing a bridge for crewed space flight on future launch vehicles. The Next Generation Launch Technology program continues technology work on future launch systems, paced to address key issues that are still open, including requirements definition, level of DOD cooperation, and technical maturity. We believe this approach will provide a more resilient bridge to providing a future launch system while assuring access to space during a future transition from the Shuttle.

Q2. If the Shuttle orbiters had lived up to their original design intent of one launch per week, they would have exceeded their design limits of 100 flights. Space Shuttle Discovery has flown the most missions of all orbiters at 30.

Q2a. Do you still believe the orbiters are capable of 100 flights per their original design?

Q2b. If not, how can we tell if any of them are capable of more than 30 flights?

A2a&b. It is important to note that the 100-flight design limit refers to the Space Shuttle orbiter airframes. NASA has not altered the expectation that the orbiter' airframes are capable of being flown at least 100 times with the specified maintenance and periodic upgrades to eliminate obsolescence issues associated with other Shuttle sub-systems. The Shuttle is subjected to rigorous inspection and maintenance following each flight. Each orbiter also undergoes a major overhaul every three to five years. NASA is implementing a Shuttle Service Life Extension Program to assure that the appropriate investments are made so that Shuttle can fly safely at least through the middle of the next decade.

Q3. In a Boeing impact analysis of ascent debris, six different scenarios in which tiles are lost from various parts of the wing were examined. It is my understanding this report was used to determine if NASA would proceed with Columbia*'s re-entry. After reviewing the analysis results, only four of the six scenarios contained thermal prediction results of "No Issue." The other two scenarios, which involve analysis of the lower wing area and the main landing gear door, has no results.*

Q3a. Can you explain why the thermal predictions under these scenarios were not summarized?

A3a. On January 23, 2003, Cases 5 and 6 were not complete prior to preparation of charts that were to be used during management reviews the following day. Case 5 was completed late in the evening and the results were discussed at the January 24, Space Shuttle Vehicle Engineering Office (SSVEO) tag up and at the STS–107 Mission Management Team (MMT) in addition to Cases 1 through 4. An action was taken from the MMT to follow up with the Case 6 results. The Case 6 results were discussed at the January 27, 2003 SSVEO tag up and the STS–107 MMT.

Q3b. Has NASA or Boeing since analyzed those scenarios and, if so, what are the results?

A3b. At the SSVEO tag up and the STS–107 MMT conference on January 24, 2003, the results for Case 5 were discussed and it was reported that the surface temperature of the analyzed areas could reach 430°F, which posed no risk for entry. At the SSVEO tag up and the STS–107 MMT conference on January 27, 2003, the results for Case 6 were discussed and it was reported that the surface temperature of the analyzed areas could reach 705°F, which could result in a localized "soft spot" in the area, but posed no safety of flight issue. This area would require post-flight inspections and repairs. The analysis did not identify any potential burn through of the structure for any of these cases.

Q4. Did NASA sacrifice funding for Shuttle safety in order to pay for cost overruns on the space station? If not, has the cost overruns on the Station program prevented increased funding for the Shuttle upgrades?

A4. No Space Shuttle safety upgrades funding has been used to supplement the International Space Station budget.

Q5. The Columbia *Accident Investigation Board, which is currently charged with investigating the* Columbia *accident, was created by NASA and reports to NASA, leading to criticism that it is not an independent body.*

Q5a. Why is this approach better than having the White House set up a blue-ribbon panel that is clearly independent of NASA, as was done after the 1986 Challenger *accident?*

A5a. Prior to *Challenger,* NASA had no contingency plan for conducting accident investigations; as a result, there was a delay before the *Challenger* investigative panel was put into place. One of the lessons learned was to have a standing panel of experts from outside the Agency to be available immediately to conduct an external investigation.

NASA's goal is to find the cause of the accident, to fix it, and to return to flight. We will support any approach that the Gehman Board, Congress, and the Administration feel best accomplishes this goal. Since all but one of the members are from organizations external to NASA, the Board is not under the control of the Agency and can therefore conduct an independent investigation. Furthermore, we have received very helpful advice and counsel from Members of Congress that the Board's charter should include revisions to strengthen the independence of the investigation, and to enable it to be as thorough as possible. NASA has been responsive to these suggestions, and has moved expeditiously to make appropriate changes to the original charter.

Q5b. Are there problems with having the Board staffed by NASA and subject to NASA disclosure rules?

A5b. The NASA employees initially assigned as staff to the Board primarily provide administrative support and have been phased out over the past month. A Task Force Team comprised of NASA employees has been created by the Board as a conduit to obtain information and data that the Board requests from NASA in the form that the Board wants. There have been modifications to the Gehman charter to strengthen the independence of the *Columbia* Accident Investigation Board (CAIB). These changes have satisfied Admiral Gehman regarding questions of independence of the Board.

Q5c. Do you have any plans for increasing the membership of the Board?

A5c. Admiral Gehman, not NASA, has already determined the need for additional Board members. Additional appointments to the Board include Dr. Sheila Widnall, Dr. Douglas Osheroff, Dr. Sally Ride, and Dr. John Logsdon. Admiral Gehman has the authority to augment the Board with whatever additional members and resources he deems necessary.

Q6. Recent news reports have charged that changes in NASA's budgeting accounts in the mid-1990s forced the Space Shuttle program to compete with the International Space Station and the X–33 program for funding. How did the need to fund the Space Station and the experimental space plane affect NASA's budgeting for long-term use of the Space Shuttle?

A6. Beginning in FY 1990, in an effort to better utilize precious resources, restructured its budget to align the Space Shuttle and Space Station programs under a single Human Space Flight appropriations account. This action allowed the Space Flight Programs the necessary flexibility to incorporate efficiencies by combining functions common to both program. Some of these activities included crew training and mission control functions.

Q7. In 2001, NASA announced that it had a $218 million shortfall in its budget for Space Shuttle operations. To solve this problem, NASA proposed canceling or delaying Space Shuttle safety upgrades, and delaying major upgrades of the Space Shuttles Discovery *and* Endeavour. *What affect did this shortfall and subsequent actions by NASA have on Space Shuttle safety?*

A7. In the FY 2002 Operating Plan, the Space Shuttle program cancelled or deferred several upgrades because of cost growth or technical immaturity. In the Operating Plan, reviewed by Congress, the funding made available as a result of these actions was then applied to Space Shuttle operations to accommodate operations cost growth. These actions did not affect safety.

Q8. A February 3, 2003, article in the New York Times *alleges that NASA removed five of the nine members of the Aerospace Safety Advisory Committee, who warned that work on long-term Shuttle safety "had deteriorated." According to the report, NASA claimed that it had changed the charter of the group. Could you please explain NASA's rationale for changing the charter of this committee and what factors led to this change?*

A8. Based on the recommendations of the Shuttle Independent Assessment Team (SIAT) and the Inspector General, in April 2002, NASA requested that several ASAP Members and Consultants step down from their positions to make room for new members with a different and more current skill mix. The Agency deemed this necessary to better reflect the current demands of its programs. All of those asked to step down had served on the Panel for at least six years—the normal term of a panel member. The average tenure of those asked to step down was 12 years.

Q9. There has been some discussion as to whether or not the re-entry path for the Columbia *could have been altered to reduce the heat build-up in certain parts of the orbiter. Could* Columbia's *re-entry path have been altered to reduce heat to the left wing?*

A9. The CAIB is attempting to determine the cause of the *Columbia* accident. While we still do not know what caused the accident, it is unlikely that the thermal profile could have been significantly reduced. The re-entry path is already optimized to minimize heating.

Q10. The public has asked why the astronauts were unable to eject from the Columbia *as it broke apart. In 2001, NASA spent $5 million to study crew escape systems, such as ejections seats and a detachable cockpit that could fly away as an escape pod and float down with parachutes. What were the results of this study, and are such escape systems practicable?*

A10. NASA has continued to study crew escape systems, including systems during re-entry. Over the past several years, a series of studies on the subject were commissioned as a part of NASA's Shuttle upgrades analysis.

Various concepts for crew escape systems have been considered against critical operating parameters. Some considerations which limit potential crew escape systems include: vehicle structure, weight, operational use, and thermal environment.

It is highly unlikely that any of the proposed crew escape concepts NASA has considered would have allowed the crew to safely escape in the dynamic environment we believe the Space Shuttle *Columbia* experienced. They were far too high and going far too fast for the existing escape system.

Currently, the astronauts wear parachutes and have a pole deployable from the crew hatch that allows the crew to bail out in level, sub-sonic flight at 25,000 feet or below.

Q11. Last December, RAND completed a study looking at possible privatization scenarios for the Shuttle program. The FY 2004 budget appears to indicate that NASA intends to pursue privatization alternatives. Has this tragic accident changed your view of expanding Shuttle privatization?

A11. The operational work of the Space Shuttle program has always been primarily performed by a contractor workforce. However, the Government has provided insight, oversight and technical expertise.

As of this date, NASA has exercised a 2-year extension option to the current SFOC contract, which carries the contract through October 2004. NASA's FY 2004 budget does not provide for "privatization alternatives," but rather assumes continued exploration of alternatives for competitive sourcing of Space Shuttle, flight operations. Further examination of Shuttle competitive sourcing options is being held in abeyance until the Gehman Board recommendations are received and assessed. It would be premature for NASA to propose any detailed plans for Shuttle competitive sourcing prior to receipt of the *Columbia* Accident Investigation Board conclusions.

Q12. Based on your plans prior to the loss of the Columbia *orbiter, I understand that NASA was planning to fly that orbiter in November to support continued construction of the International Space Station (ISS). Attainment of "core complete" was projected for around February 2004. Because of the Shuttle fleet's major modifications schedule, I understand that only three orbiters would have been available at any given point in time. For example, Discovery is now undergoing major modifications and is not scheduled for a mission until July 2004.*

Q12a. In light of recent events, do you envision any major changes to the Shuttle modifications and ISS assembly schedule when flights resume?

A12a. It is premature for NASA to predict any potential changes to Shuttle vehicle or modification schedule. We will wait until the Gehman Board has completed its investigation, found the cause of the accident and provided its findings. Although NASA can meet both ISS assembly and other agency science priorities with the three remaining Orbiters, the schedule for these missions is under review.

Q13. In his testimony before the Commerce Committee on September 6, 2001, William Readdy, the then-Deputy Associate Administrator of the Office of Space Flight, stated that "[e]very single Space Shuttle employee is empowered to call a 'time out' if they believe that there may be a potential threat to safety."

Q13a. Could you please describe process for a Space Shuttle employee to call a "time out?"

Q13b. Does this ability apply to only NASA employees or also include contractors?

A13a,b. United Space Alliance has a formal Time-Out Policy (E–02–18) signed by the Vice President, Safety Quality & Mission Assurance, that encourages and actively supports the safety practice of calling a "time out" when anyone is unsure or uncomfortable with any situation.

The term "time out" is used by United Space Alliance for its employees. However, policies are in place for all employees, whether civil service or contractors to stop any activity that they feel is unsafe. This safety awareness behavior is highly encouraged and rewarded at all levels.

Q13c. Could a "time out" be applied to a landing operation?

A13c. "Time out" is a term that is used in the ground processing of the Space Shuttle vehicle. During mission operations, all technical issues are documented and processed through the Mission Management Team. However, there are corollary policies in place for all phases of the Shuttle's mission that allow employees to raise safety issues.

Q13d. Did any NASA employee or contractor attempt to call a "time out" before or during the Columbia *mission?*

A13d. No "time out" was called during the *Columbia* mission. During the mission, following examination of launch film showing a debris hit on the orbiter, a variety of potential safety issues were raised, discussed, and reviewed through the normal agency process. Based on these reviews, the MMT judged that there was not a safety of flight issue. The Gehman Board is reviewing all documentation related to the processing, pre-launch, and launch activities for the STS–107 mission.

Questions submitted by Senator Frank R. Lautenberg

Q1. John Macidull, a member of the presidential commission that investigated the Challenger *disaster, has raised questions as to why NASA managers did not prepare alternative strategies for* Columbia*'s re-entry, in response to the damage which occurred at liftoff.*

After you saw the video footage of the liftoff, did you suspect the loss of heat shield tiles, and did you consider alternate plans to assure a safe re-entry?

A1.

- Based on our analysis during the *Columbia*'s mission, we did not believe that there was a critical threat to the Shuttle's safety posed by the foam impact.
- Each Shuttle mission entry profile is optimized for the best re-entry approach. It is unlikely that the thermal profile could have been significantly reduced.
- Had we been aware that there was a critical problem during the *Columbia*'s mission, we would have used all means at our disposal to resolve the problem and return the crew safely to Earth.

Q2. We've learned that at the same time the Space Shuttle Program was being asked to undertake additional flights to support the International Space Station, its budget was, being reduced. You were being asked to do more with less. If your funding had not been reduced, are there additional safety upgrades you would have performed?

A2.

- NASA proposed, and received, funding for those upgrades we considered to be the highest priority, and which provided the greatest return in terms of safety improvements for the Shuttle. We did this within the framework of the President's budgets and Agency priorities.
- NASA's FY 2003 budget amendment increased outyear funding for the Space Shuttle program. This increase provides for an additional flight in support of the ISS and funding for the Shuttle Service Life Extension Program (SLEP). The amendment increases funding for upgrading the Space Shuttle system by

approximately $660 million for the FY 2004–2008 timeframe. The budget amendment recognized that the Space Shuttle would be the workhorse for Space Station transport through at least the middle of the next decade.

Q3. The Columbia *flights scheduled for August 2000 and March 2002, were both delayed due to last minute safety concerns about the tiles and insulation. What changes were made after those delays to improve the safety of the* Columbia*? Were any safety upgrades made on the Shuttle during that period?*

A3.

- Neither STS–93 nor STS–109—the two previous missions of *Columbia* (OV–102)—experienced delays due to problems with the orbiter's thermal protection system (tiles and insulation).
- STS–93 Delays:
  - The launch of STS–93 was delayed twice due to ISS manifest problems first to December 3, 1998 and then to January 21, 1999.
  - The launch date was moved seven more times due to delays in the readiness of the Chandra satellite and the inertial upper stage booster.
  - The launch count down for STS–93 was aborted twice, first for a concern with hydrogen concentrations in the aft of the Orbiter; the second time was due to trans-oceanic abort landing site weather issues.
  - STS–93 launched on July 23, 1999.
  - No launch delays were caused by issues with the Orbiter's thermal protection system.
  - No new hardware or upgrade was required prior to launch.
- Between STS–93 and STS–109 *Columbia* underwent its scheduled year-long structural inspection, major maintenance and modification. During this time a number of upgrades were installed to combat obsolescence and improve safety including:
  - Multifunction Electronic Display System or "glass cockpit"
  - Micrometeoroid protection on wing leading edge and payload bay radiator doors
  - Device Driver Units—improved avionics
  - Complete wiring inspection and repair
  - Main Propulsion System upgrade of helium check valve and hydrogen fill and drain line
- STS–109 Delays:
  - STS–109 was originally scheduled for launch on November 1, 2001.
  - The launch date was rescheduled three times due to ISS manifest priorities and in delays in the OV–102 wiring inspections.
  - Two further delays were caused by problems with Hubble Space Telescope support hardware.
  - The STS–109 Flight Readiness Review established a new launch date of February 28, 2002.
  - The launch was delayed by one day due to predicted weather at KSC. STS–109 was launched on March 1, 2002.
  - No launch delays were caused by issues with the Orbiter's thermal protection system.

Q4. It appears that each time NASA requested additional funding for safety upgrades, Congress appropriated these funds. So, tell me, why didn't NASA request whatever funding was needed for all possible safety improvements?

A4.

- NASA proposed, and received, funding for those upgrades we considered to be the highest priority, and which provided the greatest return in terms of safety improvements for the Shuttle. We did this within the framework of the President's budgets and Agency priorities.
- In the FY 2002 Operating Plan, the Space Shuttle program cancelled or deferred several upgrades because of cost growth or technical immaturity. In the Operating Plan, reviewed by Congress, the funding made available as a

result of these actions was then applied to Space Shuttle operations to accommodate operations cost growth. These actions did not affect safety.

- NASA's FY 2003 budget amendment increased outyear funding for the Space Shuttle program. This increase provides for an additional flight in support of the ISS and funding for the Shuttle Service Life Extension Program (SLEP). The amendment increases funding for upgrading the Space Shuttle system by approximately $660 million for the FY 2004–2008 timeframe. The budget amendment recognized that the Space Shuttle would be the workhorse for International Space Station transport through at least the middle of the next decade. SLEP will be coordinated with NASA's other space transportation investments through the Agency's Integrated Space Transportation Plan and will make prioritized investments in the Shuttle system to ensure that the Shuttle can fly safely through at least the middle of the next decade.
- The President's budget for FY 2004 reflects our commitment to the SLEP investment process, including upgrades, necessary to safely operate the Shuttle through at least the middle of the next decade.

Questions submitted by Senator Bill Nelson

Q1. If Columbia *was indeed damaged during ascent, and this had been known by folks on the ground, what could NASA or USA have done, if anything, to ensure the safe return of* Columbia's *crew, such as changing the re-entry profile or launching a rescue mission with another orbiter?*

A1.

- NASA was unaware of any potential problems during ascent. It was only after film and video review the day following the launch that NASA became aware of potential damage to the Shuttle's tiles from debris.
- NASA has tried to develop procedures to allow on-orbit tile repairs in the past. With our current capabilities, it is not technically feasible to do so. Currently, we have no way to reach the underside of the orbiter without significant risk to the crew and Orbiter. We will be revisiting this problem in the near future.
- The orbiter did not have sufficient fuel to rendezvous with the ISS, nor did it have the required hardware to do so.
- Columbia did not have sufficient consumables to sustain the crew on orbit for more than an additional four days. The time required to prepare another orbiter for launch and train the crew for the mission would not have allowed us to launch in time to rescue the *Columbia*'s crew.
- We do not have any procedures in place that could accomplish an on-orbit transfer of crew between vehicles without compatible docking mechanisms.
- The re-entry profile for each mission is already optimized for landing with the minimum heating.
- However, had we been aware that there was a critical problem during the *Columbia*'s mission, we would have used all means at our disposal to resolve the problem and return the crew safely to Earth.

Q2. Please explain NASA's basic goals for human space flight prior to February 1. What changes have occurred in this plan, so far, since the tragic events of February 1?

A2.

- NASA's basic human space flight goals are unchanged since we articulated them in our Strategic Plan. Our first priority is, and will continue to be to ensure the safety of the public, our employees, and our high value assets such as the Shuttle.
- Space flight, both human and robotic, is the fundamental enabling capability for NASA's mission: to understand and protect our home planet, to explore the universe and search for life, and to inspire the next generation of explorers.
- Our Space Flight Enterprise goals enable this mission by:
  - Ensuring the provision of space access and improving it by increasing safety, reliability, and affordability;
  - Extending the duration and boundaries of human space flight to create new opportunities for exploration and discovery; and

○ Enabling revolutionary capabilities through new technology.

Q3. *How did NASA respond, specifically, to concerns expressed over the past several years by the Aerospace Safety Advisory Panel and others that the Shuttle program was under stress due to funding and workforce constraints? Please list dates of policy changes along with any changes in funding or workforce.*

A3.

To respond adequately to this question, one needs to examine the actions that led to the downsizing of the NASA and contractor workforce.

In August 1994, the Associate Administrator for Space Flight initiated a Shuttle Functional Workforce Review to identify the minimum workforce necessary to safely fly seven Shuttle flights per year and to establish a baseline and implementation plan to optimize the Shuttle workforce while ensuring safety.

Also in March 1994, the findings of the Space Shuttle Management Independent Review Team, under the chairmanship of Dr. Christopher Kraft, recommended that NASA consolidate all program operations under a single business entity, such as a single prime contractor.

The fundamental premise of this recommendation was:

— Separation of the operations functions from the development activities within the program by moving the government from an oversight role to an insight role.

— Elevation of the NASA-contractor interface to the program level to disengage NASA daily operations activities and empower the contractor to assume this responsibility.

— Development of a contract structure to incentivize the contractor to reduce operations costs by implementing efficiencies while maintaining safety of flight and mission success.

The findings of both the Shuttle Functional Workforce Review and the Space Shuttle Management Independent Review Team were incorporated into NASA's Zero Base Review. The objective of the Zero Base Review was to examine the NASA-wide organization and management structure, requirements and functional interfaces with a view toward moving through restructuring the reductions anticipated in NASA's out-year budget.

To respond to the reductions in the out year budget, NASA offered incentives for retirements and early-out opportunities across the Agency. In addition, the Space Flight Operations Contract with United Space Alliance, (a joint venture between Rockwell International (now Boeing) and Lockheed Martin Corporation) was initiated in October 1996 and was designed to begin a transition of Space Shuttle operations that could eventually lead to privatization. Consolidation of contracts and moving day-to-day routine Shuttle operations to the contractor while maintaining insight/oversight allowed NASA to reduce the civil service workforce and subsequently reduce cost. A review was conducted by the ASAP in the 1996/97 time frame to assess any safety impact on the initial transition to the Space Flight Operations Contract. The Panel determined that safety would not be compromised, however, they would continue to monitor the transition process. Whenever there were any indications of workforce stress observed by NASA management or anticipated gaps in skills, management responded by providing the resources necessary to ensure continued safe operations of the Shuttle.

Contractor and civil service staffing reductions continued in 1998 and 1999, mostly through attrition. In 1998 NASA's Associate Administrator for Safety and Mission Assurance verified that the staffing reduction process used by United Space Alliance (USA) the prime contractor for Space Shuttle operations, did not compromise safety. Also, a subgroup of the Aerospace Safety Advisory Panel (ASAP) observed the review by the Risk Assessment Team and provided concurrence with the Team's findings.

In the FY 2000 (CY 99) budget process the Office of Space Flight (OSF) was provided relief from the downsizing. Also, in 1999 the Agency conducted a NASA-wide Core Capability Assessment, a center-by-center analysis to identify workforce and infrastructure requirements. One of the objectives of the review was to help chart a strategy that would provide the OSF Centers with the requisite flexibility to attract and retain the critical skills necessary to ensure safe mission and program success.

The ASAP Report (published in February 2000) noted that the effects of the hiring freeze and downsizing had produced critical skills deficits in some areas and growing workforce pressures. The Panel applauded NASA's rehiring efforts but recommended that the Agency should continue to aggressively address workforce issues.

The aging of our workforce placed a strong emphasis on entry level recruiting. It naturally follows that many of NASA's critical specialty and program vacancies were filled from within and then these newly created vacancies were filled with fresh outs. Therefore, while the ability to hire allowed us to fill many critical needs, new hires did not often go directly to critical vacancies but rather to replace employees who had moved internally to where we needed them the most.

Over the past several years, the workforce hiring in the Space Shuttle program has helped to address critical hiring needs. While new hires addressed needs in understaffed critical skilled areas, workload levels have also increased. Due to the reduced number of experienced Science and Engineering (S&E) personnel in the current pipeline, we have to maintain the current level of program support and we have to ensure that our new S&E hires receive the proper skill training to meet future program needs. Given our S&E retirement eligible pool, a competitive job market for technical skilled personnel and a reduced pool of S&E graduates, we will have to enhance current recruitment and retention strategies.

From an Agency perspective, NASA uses a number of tools and flexibilities to recruit and retain critically needed skills. For example, the Agency offers starting salaries above the minimum rate, when necessary, and offers recruitment bonuses and retention allowances, when appropriate. NASA has also established a National Recruitment Initiative to develop Agency-wide recruitment strategies to attract and hire a highly technical S&E workforce. In addition, NASA will continue to use programs, such as the Presidential Management Intern Program, the Co-operative Education Program, and the Federal Career Intern Program as sources for entry level hires. The Agency may also repay student loans to attract or retain employees in critical positions.

The GAO reviewed NASA's use of Human Capital in early CY 2000 and determined that the actions NASA was taking were sufficient to sustain a quality workforce.

It was noted in the 2001 ASAP Report that although workforce concerns continued to be a focus of the Panel, they saw no safety shortfall attributable to workforce issues.

NASA has continued to provide updates to the GAO regarding progress in the workforce efforts. As part of the FY 2003 budget process, NASA conducted a Strategic Resources Review. This review highlighted several areas for transforming current business operations.

In addition, consistent with the President's Management Agenda initiative on Strategic Management of Human Capital and to support Agency SRR activities and decisions, the Office of Human Resources and Education's Functional Leadership Plan, dated May 2000 (which was developed after consultation with the Enterprises), is now being revised to reflect a more comprehensive, agile Agency-wide human capital strategic plan. As decisions are made, the Agency will address each change using the plan as a framework.

Our attention to workforce issues is focused on the future as well as the present. Recognizing that veteran space flight program employees will be retiring in the coming years, we have been aggressively taking action to attract and retain a high performing workforce to replace them.

We have a complement of skilled and dedicated civil servants and contractors who are fully able to perform the work required to ensure the continued safety and viability of our space program.

Q4. What strategy should guide operation of the International Space Station while the Space Shuttle system is grounded? Should permanent occupancy of the space station be suspended until the Shuttle system is operating again, or should the space station partners rely on Russian Soyuz and Progress spacecraft to bring crews and cargo to space station?

A4.

- The ISS continues to function well and the Expedition 6 crew continues to perform science and routine ISS maintenance.
- It is important to keep the ISS crewed. Without routine maintenance by an onboard crew, reactivating ISS after re-crewing becomes increasingly more complicated. There is also some risk associated with a de-crewed ISS. System failures are best handled by the on-orbit crew and some failures require crew interaction to resolve. Crewing also continues the on-going science activities.
- There are no threats to the ISS or the crew in the near-term and we are working options with our International Partners to be able to sustain both

until we return to flight. The crew can remain on the ISS through at least June 2003 with the logistics delivered on the Progress that docked February 4. We are evaluating options for the continued crewing and operations of the ISS beyond this time period.

- Under the current planning with our International Partners, the Expedition 6 crew of three will return on Soyuz 5S, which is currently docked to ISS as the emergency return vehicle, in late April or early May and will be replaced by a crew of two on Soyuz 6S, the replacement rescue vehicle. The planned three Progress cargo spacecraft in 2003 will be increased to four and the four Progress in 2004 will be increased to five to provide sufficient logistics to sustain the crew. With these changes and the continued rotation of a two person crews on future Soyuz replacement missions, plans are in place to keep the ISS continuously crewed through 2004.
- In the unlikely event that de-crewing is required, the ISS can be configured and de-crewed—as a contingency procedure. The ISS can remain without a crew for an extended period of time while maintaining altitude with unassisted Progress re-boost. Progresses routinely dock to and re-boost the ISS without crew interaction.

Q5. If the decision is made to rely on Russian Soyuz and Progress spacecraft beyond those that Russia already has agreed to provide at no cost to the other partners, who will pay for them. How do these plans take into account the requirements of the Iran Nonproliferation Act (P.L. 106–178), which prohibit NASA from making payments to Russia, in cash or in kind, in connection with the space station program unless the President certifies to Congress that Russia is not proliferating nuclear or missile technologies to Iran?

A5.

- NASA has met with the Russians and our other International Partners and has briefed them on the status of the ISS and the *Columbia* investigation. The maintenance of the ISS is a Partnership issue and will be resolved in this cooperative framework.
- Russia is capable and willing to provide one additional Progress in both 2003 and 2004 to sustain the ISS crew. However, Russia has indicated that doing so will require additional partner funding. The International Partners are working together in an attempt to resolve this issue.
- Any arrangements reached with the Russians or any other Partner will be in compliance with existing U.S. law and policy.

Q6. The Columbia *tragedy and the subsequent grounding of the orbiters may have a significant impact on the core capabilities supporting the Space Shuttle and the ISS through the loss of key personnel with unique technical expertise by corporate layoffs as well as with the loss of unique assets such as the orbiter and research modules. What immediate impact will the* Columbia *tragedy have on NASA's workforce or field centers? How does NASA recover from this loss? How do these critical assets, both human and hardware, get replaced once you have identified the cause, fix the problem and get back to flying again? Is it possible to estimate what impact it will have on future workforce needs?*

A6.

- NASA has established a program to help all NASA, NASA contractors and NASA grantees and their families to cope with any consequences of this disaster through its Employee Assistance Programs. NASA has placed counselors, expert in trauma and critical incident stress management in the field to help workers directly. NASA will continue to monitor this issue and do everything possible to minimize the possibility of PTSD.
- We do not anticipate any reductions in the work force as a result of the *Columbia* accident at this time.
- The Space Shuttle workforce is deeply dedicated and committed to flying the Space Shuttle safely. Currently, many of our highly skilled employees, both civil service and contractor, are supporting the efforts of the Gehman Board. Others are performing their regular duties and will continue to do so until the cause of the accident is determined and any recommendations from the Board are implemented.
- Although we do not know exactly how long it will be until we return to flight, there have been several extended periods of time between Space Shuttle launches in the twenty-two years of Space Shuttle operations. After the *Chal-*

lenger accident, the Space Shuttle workforce was dedicated to implementing improved processes and procedures that greatly enhanced the way the Shuttle vehicle was prepared, tested and validated for flight.

- Also, after the *Challenger* accident and during the 1990s there were several technical issues causing a "stand down" of the Shuttle program lasting six or more months. During those extended periods of time, the workforce kept their skills well honed by participating in simulations and extensive training in addition to performing many activities to catch up on a backlog of activities, such as facility maintenance.

Question submitted by Senator John B. Breaux

Q1. Given the expended period of time that is expected before another Shuttle flight is undertaken, what steps are you taking to ensure the Shuttle technical base and manpower doesn't suffer during this down time so that we aren't behind the eight ball when we start up again?

A1. The Space Shuttle workforce is deeply dedicated and committed to flying the Space Shuttle safely. Currently, many of our highly skilled employees, both civil service and contractor, are supporting the efforts of the Gehman Board. Others are performing their regular duties and will continue to do so until the cause of the accident is determined and any recommendations from the Board are implemented.

Although we do not know exactly how long it will be until we return to flight, there have been several extended periods of time between Space Shuttle launches in the twenty-two years of Space Shuttle operations. After the *Challenger* accident, the Space Shuttle workforce was dedicated to implementing improved processes and procedures that greatly enhanced the way the Shuttle vehicle was prepared, tested and validated for flight.

Also, after the *Challenger* accident and during the 1990s there were several technical issues causing a "stand down" of the Shuttle program lasting six or more months. During those extended periods of time, the workforce kept their skills well honed by participating in simulations and extensive training in addition to performing many activities to catch up on a backlog of activities, such as facility maintenance.

Appendix 2:

Additional Material for the Record

NATIONAL AERONAUTICS AND SPACE ADMINISTRATION
CHARTER OF THE
AEROSPACE SAFETY ADVISORY PANEL

1.0 PURPOSE

This charter sets forth the authority for, and the duties, procedures, organization, and support of the Aerospace Safety Advisory Panel.

2.0 AUTHORITY

2.1 Establishment: The Aerospace Safety Advisory Panel (hereafter called the Panel) was established as an advisory committee under Section 6 of the National Aeronautics and Space Administration Authorization Act, 1968, as amended (Pub. L. 90-67, codified at 42 U.S.C. § 2477). The NASA Administrator, having determined that it is in the public interest in connection with the performance of Agency duties under the law and with the concurrence of the General Services Administration, hereby renews the Panel pursuant to the Federal Advisory Committee Act (FACA), 5 U.S.C. App. §§ 1 et seq.

2.2 Other Statutory Authorities: 5 U.S.C. Appendix, The Federal Advisory Committee Act (Pub. L. 92-463). 5 U.S.C. § 3109, Employment of Experts and Consultants.

2.3 Executive orders: Executive Order 12024, dated December 1, 1977. Executive Order 12838 (58 FR8207, 2/12/93) dated February 10, 1993.

3.0 DUTIES

3.1 Statutory Duties: The duties of the Panel are set forth in 42 U.S.C. § 2477, as follows:

"The Panel shall review safety studies and operations plans referred to it and shall make reports thereon, shall advise the Administrator with respect to the hazards of proposed or existing facilities and proposed operations and with respect to the adequacy of proposed or existing safety standards and shall perform such other duties as the Administrator may request."

3.2 Specified Duties: To perform their tasks, members of the Panel and Panel consultants must become familiar with relevant NASA plans, policies, and guidelines including: NPD 1001.1, NASA Strategic Plan; NPG 1000.2, Strategic Management Handbook; applicable Center Implementation Plans; applicable NASA Policy Directives/Guidelines; and NASA's Agency Safety Initiative (ASI) which provides guidelines on safety for the public, for astronauts and pilots, for employees, and for high-value equipment. Pursuant to carrying out its statutory duties, the Panel will review, evaluate, and advise on those program activities that, in its judgment, contribute to safety risk and will provide identification and assessment of these to the NASA Administrator. Priority will be given to those programs that involve the safety of human flight. As part of its duties, the Panel may examine any element of NASA's activities that it believes involves safety, including the elements listed below:

3.2.1 Public Safety: This element includes those activities which provide protection for the well-being of people and prevention of damage to property not involved in NASA's business, but which may nevertheless be exposed to potential hazards associated with carrying out this business. Public safety activities performed within the Agency include, but are not limited to, such functions as:

1) Determination of public safety criteria.

2) Establishment and control of public safety hazards associated with facility and systems tests and operations.

3) Establishment and implementation, as required, of emergency or catastrophe control plans.

3.2.2 Astronauts and Pilots: This element includes those activities which provide protection for the well-being of NASA's Astronauts and Pilots. Astronaut and Pilot safety activities performed within the Agency include, but are not limited to, such functions as:

1) Determination of flight operations safety criteria.

2) Establishment and control of astronaut and pilot safety hazards associated with flight systems tests and operations.

3) Establishment and implementation, as required, of emergency or catastrophe control plans.

4) Establishment and implementation of safety standards and procedures for operation of program support and administrative aircraft.

3.2.3 Industrial Safety for Employees and High Value Equipment:

This element includes those activities which provide protection for the well-being of personnel and prevents damage to property involved in NASA business and exposed to potential hazards associated with carrying out this business. Industrial safety relates especially to the operation of facilities in programs of research, development, manufacture, test, operation, and maintenance. Industrial safety performed within the Agency includes, but is not limited to, such functions as:

1) Determination of industrial safety criteria that protect employees and High-Value Equipment.

2) Establishment and implementation of safety standards and procedures for operation and maintenance of facilities, especially test and hazardous environment facilities.

3) Development of safety requirements for the design of new facilities.

3.2.4 Systems Safety: This element includes those activities specifically organized to deal with the potential hazards of complex R&D systems that involve highly specialized areas of technology. It places particular emphasis on achieving safe operation of these systems over their life cycles, and it covers major systems for aeronautical and space flight activities, including associated ground-based research, development, manufacturing, and test activities. Systems safety activities performed within the Agency include, but are not limited to, such functions as:

1) Determination of systems safety criteria, including criteria for crew safety.

2) Performance of systems safety analyses.

3) Establishment and implementation of systems safety plans.

3.2.5 Safety Management: This element includes both the program and functional organizations of NASA and its contractors involved in the identification of potential hazards and their elimination or control, as set forth in the foregoing description of safety activities. It also includes the management systems for planning, implementing, coordinating, and controlling these activities. These management systems include, but are not limited to, the following:

1) The authorities, responsibilities, and working relationships of the organizations involved in safety activities, and the assessment of their effectiveness.

2) The procedures for insuring the currency and continuity of safety activities, especially systems safety activities which may extend over long periods of time and where management responsibilities are transferred during the life cycles of the systems.

3) The plans and procedures for accident/incident investigations, including those for the followup on corrective actions and the feedback of accident/incident information to other involved or interested organizations.

4) The analysis and dissemination of safety data.

4.0 PROCEDURES

4.1 Panel and NASA Administrator: The Panel will report to and function in an advisory capacity to the NASA Administrator.

4.2 Information Gathering and Reviews: The Panel will be provided with information necessary to discharge its advisory responsibilities which pertains to both NASA and its contractors' safety activities. This information will be made available through the mechanism of appropriate reports and by means of in situ reviews of safety activities at the various NASA and contractor sites, as deemed necessary by the Panel and arranged through the Aerospace Safety Advisory Panel Executive Director for the Administrator. The Panel shall send observers to NASA safety reviews to the maximum extent possible including: Prelaunch Assessment Reviews/Joint Prelaunch Assessment Reviews (PAR's/JPAR's) and Flight Readiness Reviews/Joint Flight Readiness Reviews (FRR's/JFRR's).

5.0 ORGANIZATION

5.1 Membership:

5.1.1 Panel Members: The Panel will consist of a maximum of nine members who will be appointed by the NASA Administrator. Appointments will be for a term of 6 years. The NASA Administrator may, at his/her discretion, choose to extend a member's term by no more than 6 years beyond that member's original term limit.

5.1.2 Consultants: The NASA Administrator may appoint consultants for a specific task at the request of the Panel. A consultant must have expert knowledge for the specific task assigned. A consultant shall serve the lesser of 1 year or the time required to complete the effort defined by the Panel Chairperson.

5.1.3 NASA Membership: Not more than four members of the Panel can be current employees of NASA, nor shall NASA members constitute a majority of the composition of the Panel and its consultants at any given time.

5.1.4 Ex Officio Membership: The NASA Headquarters Associate Administrator for Safety and Mission Assurance shall serve as the sole Ex Officio member of the Panel. In this capacity, this member may participate in the Panel's activities but not as a voting member.

5.1.5 Membership Selection Process: The Executive Director shall maintain a database of candidate members and consultants for the Panel. The process of adding a member to the Panel is described in "Appointment of New Aerospace Safety Advisory Panel (ASAP) Members," HOWI 3300-Q025.

5.1.6 Member's Expertise: Membership of the Panel shall be balanced fairly in terms of expert knowledge, professional and/or academic background, diversity, and recent professional experience in their area of technical expertise.

5.2 Officers:

5.2.1 Chairperson/Vice Chairperson: The officers of the Panel shall be a Chairperson and a Vice Chairperson who shall be selected by the Panel from their membership to serve for a 2-year term which may be extended with the written approval of the NASA Administrator.

5.2.2 Powers: The Chairperson or Vice Chairperson in his /her absence shall have the usual powers of a presiding officer. The Chairperson may organize the Panel as he/she deems appropriate to accomplish the Panel's statutory duties.

5.3 Meetings: Regular meetings of the Panel will be held as often as necessary and at least once a year. One meeting each year shall be designated as the Annual Meeting. The purpose of the Annual Meeting is to present the Panel's Annual Report to the Administrator and the Public. The Chairperson will propose the meeting agenda, to be approved by the Executive Director, and will officiate at the Annual Meeting.

5.4 Reports and Records:

5.4.1 Annual Report: The Panel shall submit an annual report to the Administrator. Instructions on this report's preparation may be found in "Prepare Aerospace Safety Advisory Panel (ASAP) Annual Report," HOWI 1156-Q026.

5.4.2 Safety Reviews and Evaluations: The Panel will submit to the Administrator, at his/her request, reports and advice required by statute.

5.4.3 Congressional Correspondence/Testimony: The Panel may be requested to provide written responses to congressional inquiries and/or testimony before Congress on specific NASA safety issues.

5.4.4 Files/Records: All records and files of the Panel, including agendas, minutes of Panel meetings, studies, analyses, reports, or other data compilations or working papers, made available to or prepared by or for the Panel, will be retained by the Panel.

5.4.5 Minutes of Meetings: Minutes of all meetings of the Panel established by the Panel will be kept. Such minutes shall, at a minimum, contain a record of persons present, a description of matters discussed and conclusions reached, and copies of all reports received, issued, or approved by the Panel. The accuracy of all minutes will be certified by the Chairperson (or by the Vice Chairperson in his/her absence).

5.5 Compensation: Members of the Panel and Panel consultants will be compensated for their services in accordance 42 U.S.C. § 2477 and will be allowed travel and per diem expenses as authorized by 5 U.S.C. § 5701.

5.6 Avoidance of Conflicts of Interest:

5.6.1 Nongovernmental Members: Nongovernmental members of the Panel and Panel consultants are "Special Government Employees" within the meaning of 18 U.S.C. § 202.

5.6.2 Financial Disclosure: Nongovernmental members of the Panel and Panel consultants will submit an "Executive Branch Confidential Financial Disclosure Report" (Form 450) prior to participating in the activities of the Panel or acting as a Panel consultant.

6.0 SUPPORT

The NASA Headquarters Office of Safety and Mission Assurance will provide a staff, to be comprised of full-time NASA employees, to provide support to the Panel. The members of this staff shall be fully responsive to direction from the Chairperson of the Panel.

6.1 Executive Director: The Executive Director is appointed by the NASA Administrator, is the Designated Federal Official, and will serve as the Executive Secretary and Technical Assistant to the Panel. The Executive Director shall have the responsibility for: agenda approval, administration of the affairs of the Panel, general supervision of all arrangements for safety reviews and evaluations, maintaining a written record of all transactions, Panel meeting minutes, and trip reports.

6.2 Staff Assistant: The Panel's Staff Assistant shall be responsible for general staff support for the Executive Director, the coordination of personnel and legal forms for the Panel, and other duties as assigned.

6.3 Secretary: The Panel's secretary shall be responsible for general administrative support for the Panel members, Staff Assistant, and Executive Director. The secretary will provide correspondence support, travel support, schedule maintenance and support, and support for personnel and payroll forms.

6.4 Budget: The NASA Headquarters Office of Safety and Mission Assurance will provide the budget for operation of the Panel. The estimated annual operating cost totals $555,000, including 3.0 workyears for staff support.

7.0 CHARTER RENEWAL

Pursuant to the Federal Advisory Committee Act, 5 U.S.C. App., this charter expires 2 years from approval date and can be renewed if the NASA Administrator determines that it is in the public interest in connection with the performance of Agency duties under the law and with the concurrence of the General Services Administration. This charter replaces the Charter of the NASA Aerospace Safety Advisory Panel, dated May 1, 2001.

Administrator

May 1, 2003
Date